The Cultural Interpretation of the Brick Carvings of Bozhou Gorgeous Dramatic Stage

亳州花戏楼砖雕艺术文化解读（中英文）

唐利平 著

北京师范大学出版集团
BEIJING NORMAL UNIVERSITY PUBLISHING GROUP
安 徽 大 学 出 版 社

图书在版编目(CIP)数据

亳州花戏楼砖雕艺术文化解读:汉英对照/唐利平著.
—合肥:安徽大学出版社,2014.7
ISBN 978-7-5664-0781-8

Ⅰ.①亳… Ⅱ.①唐… Ⅲ.①古建筑—砖—装饰雕塑—建筑艺术—亳州市—汉、英 Ⅳ.①TU-852

中国版本图书馆CIP数据核字(2014)第135900号

亳州花戏楼砖雕艺术文化解读(中英文) 唐利平 著

出版发行:北京师范大学出版集团
安 徽 大 学 出 版 社
(安徽省合肥市肥西路3号 邮编230039)
www.bnupg.com.cn
www.ahupress.com.cn
印 刷:中国科学技术大学印刷厂
经 销:全国新华书店
开 本:170mm×230mm
印 张:11.75
字 数:220千字
版 次:2014年7月第1版
印 次:2014年7月第1次印刷
定 价:29.00元
ISBN 978-7-5664-0781-8

策划编辑:李 梅 钱来娥
责任编辑:钱来娥 李 雪
责任校对:程中业
装帧设计:李 军 金伶智
美术编辑:李 军
责任印制:赵明炎

序言

砖雕是一种中国传统建筑装饰艺术，与木雕、石雕合称为中国建筑三雕。砖雕取材便利，题材广泛、内容丰富。砖雕艺术品清新质朴、造型典雅，不仅有很强的艺术效果，还体现了艺术家的审美情趣和愿望诉求，传递着文字语言难以表达的深层文化特质。从宋代起，砖雕成为中国上至皇宫、下至民间普遍流行的建筑装饰。时至今日，中国各地随处可见古朴精美的砖雕作品点缀着各类古建筑。其丰富的表现内容和深刻的文化内涵形成了独树一帜的建筑装饰艺术，被列入国家级非物质文化遗产名录，在中外建筑艺术中久负盛誉。中国安徽省亳州市区城北关的花戏楼（建于清代顺治十二年）就因砖雕作品群规模大而集中，雕工技艺精湛，雕作画面精美、内容多样，文化蕴涵丰富而闻名于世。中央电视台《国宝档案》节目称亳州清代花戏楼为“国珍”。“它的雕刻和彩绘工艺闻名于世，其中水磨砖雕更是精美绝伦”，“在不足两寸，寥寥数十平方米的水墨青砖上，共雕有52幅作品”，“方寸之地，展现给我们的却是大千世界的奇妙故事”。作者作为在花戏楼所在地亳州任教的英语学人，惊叹于古人的智慧、砖雕的精美和中华文化的博大精深，在实现“中国梦”和“中国文化走出去”的语境下，便产生了一种与他人分享砖雕之美的愿望。于是，作者开始着手研究和译介花戏楼砖雕艺术文化，目的是弘扬中华传统优秀文化，促进中外文化交流。

然而在对砖雕艺术文化的译介、研究过程中，作者发现古建的翻译不同于文学和科技类翻译，因为它是非文本翻译。它不是单纯的语言间的转化，而是先由艺术图视符号到文字符号，再由“源语”文字到“译入语”文字的转化。这种“非文本翻译”的特殊性在于：首先，因其翻译对象是砖雕图视，它缺少“源语”文本

(这种情况在许多研究中都存在,如黄山的徽州建筑文化研究)。古建以独特的建筑风格、雕刻或绘画艺术诉说着设计者或建造者赋予她的丰富文化内涵,所以它的"源语"文本是表现形式多样的建筑语言,如雕刻、彩绘、图画或图视、纹饰等艺术形式。这种非文本源语的特殊性成为了翻译工作的第一个障碍,此为译文形式之特殊。第二,"源语"砖雕图视的"形"与"质"关系密切,翻译中信息取向和诗学取向并存。中国的砖雕装饰艺术历史悠久,承载着厚重的中国文化内涵,蕴含着丰富的传统文化元素,其风格和内容是特定时代特定地域民风民俗的写照,凝聚着多元的文化思想内涵和审美情趣。许多砖雕装饰图视及其装饰源自于历史故事、民间传说或成语典故,源语形式最具诗学取向和经典内涵,其翻译的困难程度与典籍翻译的难度相仿,此为译文"内容"之特殊。第三,译文受众决定译文风格和文体。古建筑艺术的翻译属于外宣类翻译,是集艺术和各类文学作品翻译于一体的文化信息型翻译。它以向西方受众(主要是对中国古建筑艺术感兴趣的西方游客)传达艺术和文化信息为主,属于文化艺术宣传性质的翻译,译文的风格和文体需要译者认真把握和斟酌,此为译文"文体"之特殊。第四,在古建翻译文本缺失的情况下,译者同时又身兼作者的身份。"作家的连贯建构是诗学取向,译家的连贯建构是信息取向。[①]"鉴于古建筑艺术文化翻译活动的特点,译者同时又是作者。译者在整个翻译过程中首先是作者,他关注翻译信息内容的目的是进行诗学建构。这一阶段为生产"源语"翻译文本的过程,即语内的"文学创作",而后出现"语际转换"的可能性。否则,翻译将不可能实现。因此,译者的翻译过程可理解为"作者(译者本人)的诗学建构→译者的信息建构",此为译者"身份"之特殊。

针对以上列举的古建筑翻译的特殊性,我们采取了相应的翻译策略和方法。面对古建文本的缺失或内容的不完善,译者参与创建,挑战自己的文化和语言功底,这是本研究的指导思想。中国传统砖雕艺术文化的翻译要经历图视符号的解读、语内文字文本的创建,再经过"语内转换"到"语际转换"的双重过程,共三个环节。首先,对"砖雕源语"的翻译,"必须通过符号解读、符号到文字的转换和文字重建的翻译过程。译者要同时解读图视符号,将图视符号转换成"语内"(intralingua)的语言符号,并阐释图视所蕴含的文化内涵,而后进行语际符号的

---

① 彭秀林.文学翻译研究新视角——王东风《连贯与翻译》[J].中国翻译.2012(1).51.

转换和重建。更为特殊的是语内转换需要两个环节:第一,对砖雕图视的艺术解读和语内文化解读(解码),生成源语翻译文本;第二,从“源语”文本到译入语文本的“语际”(interlingua)转换和重建。如果把从砖雕图视到语言符号的转换称为“语内转换”,则语内转换显得尤为重要。所以,译者首先要“吃透”砖雕图作的文化寓意,并对此进行语内“深描”(thick description)“所谓深描,指对文化现象或符号意义进行层层深入描绘的方法”①。在对砖雕图视艺术和文化的深描过程中可对文本进行现代化的处理,以缩短砖雕图作与译者和读者的“历史距离”,创造出能让现代读者充分理解,又尽量贴近砖雕作品本意的源语文本。西方受众的审美价值观与译者不同,但他们对古建筑艺术的审美心理期待大多与译者相吻合,这样就能最大程度地通过译文读懂中国砖雕所表达的艺术形式及其所承载的文化元素,增加砖雕艺术的观赏性、凝聚性和直观性。

译文的语言风格不应该是很文气(bookish)的书面语言,但又不同于流行的通俗小说文体。在描述砖雕艺术、翻译民间故事时,文体应近似散文,娓娓道来。而对于历史故事中出现的典故或古语,如“天禄书镇”、“传鲈赐宴”、“夔一足”和“狻猊”等砖雕,就要用到“深译”(thick translation)的手法,“即通过在译文里添加注释来表现原文深厚文化语境的翻译策略”②,在保留艺术文化精髓的基础上,使用文化阐释和经典注释手段对源语文本的文化进行再阐释,并注重译文措辞的纯正性、精确性、适合性,力图帮助西方读者理解根植于中国传统文化中的砖雕艺术语言。有许多砖雕图的译名,如,“鹰扬宴”、“唐”和“一品朝”等,要对其蕴含的比喻、替代意义作阐释,否则可能会产生“译犹未译”的问题。

语言形式方面,必须考虑中西方文化在审美倾向(aesthetic preference)上的差异性。汉语讲究“语言的形象美”;英语偏好“语言的逻辑美”③。汉语文章在遣词造句时多用四字成语,讲究句式对称;汉语擅用修辞,如夸张和比喻,较喜用华丽的辞藻。而英语偏好句式中语法的逻辑性,用词的新颖独特,行文中平铺直叙的写作手法。对于外宣类作品,使用通俗的民间语言和朴实简洁的口语更合

---

① 龙吉星.当代西方翻译研究中的人类学方法初探[J].中国翻译.2013(5).6.

② 龙吉星.当代西方翻译研究中的人类学方法初探[J].中国翻译.2013(5).6.

③ 王东风.论作用于翻译过程中的跨文化因素.陈宏薇·方法·技巧·批评.翻译教学与实践研究,杨自俭、王菊泉.英汉对比与翻译研究(之八)[M].上海:上海外语教育出版社,2009:201.

适。所以,英语译文不必要使用过多的修辞手法,语言应简洁易懂,用词恰当准确,以常用词(common words)为主。行文流畅轻松,语言优美,增加砖雕艺术的感染力。最后,考虑到古建筑艺术翻译的特殊性,在解读翻译砖雕艺术作品时,译者的“深描”和“深译”能力同等重要。译者既不能是译入语文化的“局外人”,又必须成为源语文化——中国文化的行家里手。有位专家说过:“一个文化学者可以不是译者,但是一个好的译者必须同时又是文化学者。”如此,古建筑艺术文化的翻译才会成为可能。

庄子曰,“天地有大美而不言”(庄子《知北游》),而译者就是“美的发现者”。亳州花戏楼的原名是“大关帝庙”,是山西和陕西的商人(合称晋商或西商)在亳州经商时捐资建造的会馆,又名山陕会馆。建筑文化是中国文化的一部分,是社会生活和价值观的体现。建筑除了防寒避雨的居住功能外,还体现出建造者的精神追求和审美情趣。作为中国建筑常见装饰形式的砖雕是民俗与市井风情的万花筒,为我们保留了农业文明时期重要的社会文化信息①。所以,在对砖雕作品进行文化解读时,作者参阅了大量有关晋商历史和砖雕艺术的文献资料,并对现存的山陕会馆,如南阳社旗会馆、山东聊城山陕会馆、苏州全晋会馆等进行了实地调研,以便对与亳州花戏楼砖雕进行深入研究。作者试图在忠实于砖雕作品、尊重创作者的基础上,尽量深挖作品的文化内涵。汉语文本注重渲染砖雕特有的艺术魅力,正所谓“美景之美,三分在景,七分在品”。“翻译的衍生性和调节作用意味着跨文化翻译是阐释的具体化,而不是文化形式的直接转换。②”在翻译过程中,我们认为保留源语生态文化与增加译文的可读性同等重要。对砖雕作品的名称和成语典故,作者进行文化“深描”和“深译”处理,即详细的文化阐释和注释,以便增加译文的可读性和趣味性。

“译名无小事”。目前,“花戏楼”这个建筑名称有很多翻译版本,采用的方法有意译和音译。如 *China Daily* 在 2011 年 1 月 23 日对亳州的报道:“Another must-go scenic spot in Bozhou is the Huaxi Lou (Opera Theater)”,将“花戏楼”译为 the Huaxi Lou,并注释 Operate Theater。2013 年 6 月 9 日,中央电视台第四套节目(CCTV4)在《国宝档案》节目的“亳州寻珍”专题的解说词中,把“花戏楼”翻译成 The Gorgeous Dramatic Stage。另有 Huaxi Building, Flowery Stage, Huaxi

---

① 毛晓青,王彩霞. 中国传统砖雕[M]. 北京:人民美术出版社,2008. 2.

② 孙艺风. 翻译与跨文化交际策略[J]. 中国翻译. 2012(1). 20.

Lou, Hua Xi Lou 等译法。本书使用音译“the Huaxilou”和意译“the Gorgeous Dramatic Stage”两种译法。本书采用这两种译法，出于下列原因：首先，“花戏楼”指代的是一座群体建筑，而戏楼本身只是群体建筑中的一部分，所以不宜译为“Huaxi Building”。其次，花戏楼名字中有“花”字是因为它内部的精美雕刻和艳丽彩绘，而不是因为它是一座用花装饰的舞台。如果译成“Flowery Stage”容易使人产生误解。使用 Gorgeous Dramatic Stage，因 gorgeous 能体现戏楼的精美雕刻，stage 展示了它旧时作为戏台的作用。再次，从文化输出的角度，用汉语拼音让读者熟悉中国文化，接受地名、建筑物的拼音用法。“花戏楼”三个字的拼音“Huaxilou”写在一起有整体感，既符合汉语语言规律，又简洁明了，故采用之。

综上所述，本书是对亳州古建筑花戏楼砖雕装饰艺术进行解读及翻译的一次尝试。本书分为三个部分：第一部分从建筑装饰艺术的角度介绍花戏楼及其砖雕作品，并从历史的角度介绍了花戏楼建造的过程和文化背景。第二部分从文化层面探讨了砖雕作品的审美价值，解读和翻译砖雕传达的深层文化蕴涵，让读者对中国传统的儒释道文化和晋商精神有所了解。第三部分是花戏楼砖雕作品图解，介绍了 52 幅花戏楼山门上的砖雕作品。每幅作品均配有图片，并分别用中英文对图片的内容以及砖雕的寓意进行说明和阐释。

Contents

第一部分 亳州花戏楼砖雕艺术文化解读

引 言 ………………………………………………………………… 3

第一章 花戏楼砖雕艺术 ………………………………………… 3

一、花戏楼的"三绝" …………………………………………… 3

二、砖雕装饰艺术 ……………………………………………… 6

(一)中国砖雕艺术概述 ………………………………………… 6

(二)南徽北晋的砖雕风格 ……………………………………… 8

(三)花戏楼的砖雕艺术 ………………………………………… 12

三、花戏楼建造始末 …………………………………………… 15

(一)"药材之乡"的由来 ………………………………………… 15

(二)花戏楼(山陕会馆)的修建 ………………………………… 19

第二章 花戏楼砖雕文化审美蕴涵 …… 25
一、"生生不息"哲学艺术精神的观照 …… 26
二、"参天地"中"和"的文化审美意象 …… 28
三、大关帝庙承载的忠义文化 …… 32
四、刻在青砖上的中华传统文化 …… 37
(一)经世流传的儒家"孝文化" …… 37
(二)"崇儒尚文"的晋商文化 …… 40
五、《四爱图》折射的道家艺术生命精神 …… 43
六、"祥云瑞兽"艺术语言的审美文化寓意 …… 47

第二部分 The Cultural Aesthetic Implication of the Brick Carvings of Bozhou Gorgeous Dramatic Stage

Chapter 1 The Art of the Brick Carvings of Bozhou Gorgeous Dramatic Stage …… 55
Ⅰ. The Three Wonders of Bozhou Gorgeous Dramatic Stage …… 56
Ⅱ. The Decorating Art of Brick Carvings …… 58
1. A Brief Introduction of Chinese Brick Carvings …… 58
2. The Brick Carvings of Hui-style in South China and Jin-style in North China …… 62
3. The Brick Carvings of Bozhou Gorgeous Dramatic Stage …… 67
Ⅲ. The History of the Construction of Bozhou Gorgeous Dramatic Stage …… 70
1. The Background of "the Home of Chinese Herbal Medicine" …… 70
2. The Story of Building Bozhou Gorgeous Dramatic Stage …… 74

Chapter 2 The Cultural Aesthetic Connotation of the Brick Carvings of Bozhou Gorgeous Dramatic Stage
Ⅰ. The Perspective of "Everlasting" Philosophical Artistic Spirit …… 83
Ⅱ. The Cultural Aesthetics of "Can Tian Di" and "Harmony" …… 86

Ⅲ. The "Loyalty Culture" of the Grand Temple of Guan Yu …………… 91
Ⅳ. The Traditional Chinese Culture Carved on the Brick Carvings ……… 96
1. The Everlasting Chinese "Filial Piety Culture" from Confucianism …………………………………………………………… 96
2. The Business Culture of "Respecting Confucianism and Knowledge" …………………………………………………………… 99
Ⅴ. The Taoist Artistic Spirit Reflected by *The Drawings of the Four Loves* …………………………………………………………… 103
Ⅵ. The Cultural Aesthetic Message in "Auspicious Clouds and Animals" of the Brick Carvings …………………………………… 109

第三部分 亳州花戏楼砖雕故事文化图解

正门牌坊 Carvings on the Main Gate …………………………………… 117
钟楼牌坊 Carvings on the Gate of Bell Tower ……………………………… 152
鼓楼牌坊 Carvings on the Gate of Drum Tower ……………………………… 163

后 记 ………………………………………………………………………… 173

第一部分

亳州花戏楼砖雕艺术文化解读

引　言

亳州花戏楼位于安徽省亳州市区城北关，原名“大关帝庙”，现为亳州市名胜古迹，国家一级文物保护单位。花戏楼是一座群体建筑，主要包括大关帝庙（关公祠大殿）和戏楼，是明清时代山陕商人于清顺治十二年（1656 年）所建，因此又有“山陕会馆”之称。因馆内建有祭祀山西名人关羽的庙祠，主奉关帝，故初建时称为“大关帝庙”，现属于花戏楼景区。人们之所以把关帝庙称为“花戏楼”，主要是因为庙内的一座彩绘鲜艳、雕饰精美的戏楼。戏楼有两层，下层为进入院内的通道，上层是演戏用的舞台（或称歌台）。由于戏楼遍布戏文，彩绘鲜丽，花团锦簇，蕴含着典故的精美砖雕和木雕闻名于世，后人遂以“花戏楼”传呼至今。

花戏楼由明清时期两大驰名天下的商帮——山西和陕西商人（史称晋商或西商）所建，用作在亳州经营药材的联络之地，是一座集商业、娱乐、宗教、政治功能为一体的古代建筑。花戏楼历经三百多年的悠长岁月，见证了三百年晋商文化的兴衰。如今繁华散尽，只留下一座装饰精美、布满雕花的戏楼和一片雄伟壮丽的建筑群。虽然它只是一座雕花的戏楼，但是对于亳州人来说，它却雕刻着久远的历史，诉说着明清时代晋商的辉煌和古城的繁华。亳州花戏楼，它代表着一段远去的岁月，一个繁华的背影，一抹流淌的记忆，一种精深的文化。

第一章　花戏楼砖雕艺术

一、花戏楼的“三绝”

明清时代的会馆内往往都建有戏楼，与神庙殿宇、商馆会所组合形成一片古建筑群。山陕会馆作为明清时代晋商在各地的联络处所，目前现存的有 80 余座古建筑群分布在全国各地,包括京杭大运河边的山东聊城山陕会馆、河南南阳社旗山陕会馆、湖北襄阳山陕会馆、河南唐河山陕会馆、邓州山陕会馆、洛阳山陕会馆、苏州全晋会馆等,其建筑集宫殿、庙宇、商馆、民居、园林建筑之大成,既雄伟壮观、雍容华贵,又玲珑秀丽、典雅有致;既渲染了宫殿的气势和庙宇的静穆,又充满着柔美色彩和诗情画意,给人以艺术整体美的强大震撼力。无论在建筑布局、整体设计,还是装饰艺术等方面都堪称为中国古代建筑的杰作。亳州花戏楼(亳州大关帝庙)与业内专家公认享有"天下第一会馆"美誉①的社旗山陕会馆(建于1756年,原赊店山陕会馆),在1988年元月同时被国家文物保护局列为第三批全国重点文物保护单位。根据史料记载,虽然亳州花戏楼的规模小于南阳社旗会馆,但它的出现较社旗会馆早上百年。

这座古建筑群以大关帝庙为主,由张飞庙、岳飞庙、朱公书院、火神庙和粮坊会馆等组成。大关帝庙是花戏楼的主体建筑。它的正门是一座三层牌坊式仿木结构建筑,上面布满剔透玲珑的砖雕;两边有钟鼓二楼衬托,同样配有精美的砖雕装饰。进院后可看到与之相背连接的就是全国闻名遐迩的关帝庙,再往里有正殿,包括前殿、后殿,天井院内有东西看楼。整座建筑虽然面积只有3163平方米,却也布局合理、设计巧妙,戏楼雕刻得玲珑剔透,装饰的精致美观。由于会馆在当地的影响力非常大,各地商贾贵人往往把会馆作为"比富"、"炫耀"的招牌,实际上会馆已经具有了现代意义上的商业广告性质。所以建造者往往斥巨资建造、装饰会馆,使其富丽堂皇。亳州山陕会馆的建筑装饰艺术精美华丽,工艺考究,其中砖雕、铁旗杆和木雕彩绘被称为花戏楼的"三绝"。同济大学古建筑研究专家陈从周教授在考察花戏楼后感慨:"院宇宏敞,画栋朝飞,游人至此,流连忘返。余小住一周,归程回首,迟迟举步矣。②"

花戏楼的大门也就是大关帝庙的山门,是一座仿木结构的砖雕牌坊,其"砖刻门楼硕大工整,为今存少见者。其中若干刻法较朴茂者,以清乾隆时作品为多;过于繁琐者,似清末期时所加配。但以其整体而论,确为难得。③"牌坊坐北朝南,两侧有钟楼、鼓楼和拱门相衬。整个牌坊为水磨青砖雕刻,"在不足10厘米厚的水磨砖雕上,布满了精美的立体通透雕刻,有人物山景、车马城池、虫鱼鸟

---

① 赵静. 中国赊店山陕会馆[M]. 郑州:中州古籍出版社,2013.3.
② 陈从周. 亳州大关帝庙[J]. 同济大学学报. 80(2). 85.
③ 陈从周. 亳州大关帝庙[J]. 同济大学学报. 80(2). 85.

兽、林山花草、亭台楼阁。人物形象栩栩如生,亭台楼阁美轮美奂,林木花草枝繁叶茂。雕工精细,堪称一绝。"[①]

花戏楼的正门前有两根铁旗杆,它们是花戏楼的又一绝。这对旗杆高 16 米,每根重一万两千斤。铁旗杆有蟠龙绕柱,昂首翘尾,生动传神。每个旗杆还有三层挂着风铃的方斗,风起铃响,清脆悦耳。铁旗杆高大雄伟,造型独特,令世人称奇。

木雕彩绘是花戏楼的最后一绝,也是花戏楼的灵魂所在。戏台四周布满木雕彩绘,木雕内容主要是三国戏文故事。木雕中的人物形态逼真,城池山野也有很强的立体感和真实感。戏楼的木雕通体施以彩绘,色彩浓烈明快,对比强烈。木雕与彩绘巧妙结合,相辅相成,灵动鲜活。

花戏楼的砖雕、铁旗杆和木雕彩绘互为映衬、各具特色,构成了戏楼的三绝,也成为亳州山陕会馆的精华所在。除山陕会馆之外,在亳州城内留有西商印记的建筑还有南京巷钱庄、晚清"平遥帮"所开钱庄"日升号"的旧址、江宁会馆等,它们都保存得比较完好,并已被列入省级文物保护单位。

值得研究的是,文物专家对花戏楼的建筑风格和砖木二雕一直有争议。陈从周先生考证认为:"因属山陕会馆,其建筑胥出晋匠之手,一如山西所见。其与山东聊城之山陕会馆,河南开封之山陕会馆、江苏苏州之全晋会馆同出一辙。砖木二雕为该庙之精华所在,为研究清中叶建筑艺术之重要实例。[②]"而武汉理工大学的杨絮飞专家认为:"山陕会馆的正门和钟、鼓楼的正面墙采用的是标准的徽派建筑风格。山陕商人建设的花戏楼正门的风格与现代我们在徽商的故乡如宏村、西递以及婺源等地看到的建筑风格一致。只不过,花戏楼的正门比起古徽州建筑更宏大、更壮观、更精美。[③]"某些地方的山陕会馆的一些建筑构件可能来自于山陕两省,如聊城会馆的一对石狮子和社旗会馆的琉璃照壁,据讲解员介绍是在山西雕成后运至会馆。但并非只要是山陕会馆就必出自于晋匠之手。通过对河南南阳的社旗会馆、山东聊城的山陕会馆等的实地考察研究和对有关资料的分析,再将亳州花戏楼山门墙上的砖雕布局、内容形式及雕工技艺与之相比

---

① 亳州文化旅游公司. 花戏楼解说词[CD]. 2012.

② 陈从周. 亳州大关帝庙[J]. 同济大学学报. 80(2). 85.

③ 杨絮飞. 论亳州山陕会馆建筑风貌的矛盾性. 阜阳师范学院学报社会科学版[J]. 2011(3). 31.

较来看,花戏楼砖雕兼有这两种风格,因此它的建筑和装饰风格应是徽派和晋派风格的融合。可见,徽匠和晋匠的合作成就了花戏楼这一杰作。

亳州花戏楼的砖雕图视内容的选择和布局的安排也是精心独到,非常考究。亳州花戏楼的独特之处不仅在于她精美的艺术形式带给人的视觉享受,还在于她的丰富的文化内涵给人的无限想象。她把深邃的中华文化巧妙地融入砖雕艺术,以民众喜闻乐见的图视艺术形式走进人们的视野,用比喻、象征、谐音、拟人、白描和双关等手法向人们传达了博大精深的中华文化,耐人寻味,意蕴无穷,使人们在欣赏艺术中受到感染和震撼、得到训诫和警示,达到了寓教于乐的教化功能。设计者根据中华文化的思想精神和价值观念对砖雕图视内容的选择进行了精心的梳理安排,处处体现出设计者对中华传统文化的深邃领悟和很高的艺术造诣。艺术形式的对称和思想内容的对应相互结合,整体思想内容和单体图视形式安排巧妙,达到艺术形式和思想内容的完美结合和高度统一,不愧为中原大地建筑装饰艺术文化之“珍宝”。

二、砖雕装饰艺术

(一)中国砖雕艺术概述

砖雕是中国传统建筑独特的装饰艺术之一,与木雕、石雕合称为建筑三雕。在中国古代,砖雕多出现在民间建筑装饰中,是中国传统建筑装饰的重要形式之一。民间艺人运用凿和木锤,以锯、钻、刻、凿、磨等手法,在特制的、质地细密的块砖上雕刻物象或花纹,包括人物山水、花鸟动物等,然后装饰在祠堂、庙宇及民居的门楼、屋脊、角带、山墙、影壁、飞檐、栏杆等处的壁面上,起到美化建筑的作用,体现了建造者的审美情趣和愿望诉求。砖雕取材便利,且具有坚硬、耐磨和防腐等特点。从宋代起,砖雕用于官居、寺庙以及民居建筑装饰就非常盛行。砖雕通常也指用青砖雕刻而成的雕塑工艺品,图案内容丰富,人物、山水、花鸟、走兽、车马、城池、文字等无所不包。以人物为主的题材内容包括宗教神话、戏曲图谱、民俗风情、民间传说和其他社会生活等方面。砖雕技艺精湛,清新质朴,其丰富的表现内容和深刻的文化内涵体现出中华民族丰富多彩的文化和不同区域的特征,形成了独树一帜的中国传统建筑装饰艺术,体现出中国传统文化的博大精深。砖雕被列入国家级非物质文化遗产名录,在中国乃至世界建筑艺术长廊中享有美誉。

中国古建筑雕刻艺术及青砖雕刻工艺是由殷代陶艺、东周瓦当、汉代画像砖等发展演变而来的。砖雕艺术始于汉,发展于宋,鼎盛于明清。中国砖雕的形成和发展,和砖雕采用的材料是分不开的。与西方古代的石材建筑不同,中国古代的高级建筑以砖木为主要材料,民间建筑则采用土木草混合结构。砖是在陶器的发展演变中形成的。我们的祖先在八千多年以前就已经学会在陶器上装饰附加堆纹和雕刻花纹,并使用雕、塑、印、贴等多种装饰手段。早在7000多年前的新石器时代,我国古代先民就掌握了选土和烧制技术。在距今4000年左右的四坝文化遗址中发现了成型的日晒砖,由黄土加羼和料晒干而成。战国时,完烧制砖最初被作为装修材料,大量运用在墓室建筑中。秦砖素来享有盛名,目前所见的主要出土于始皇陵及周围遗址。秦砖砖色青灰,坚固耐用,制作规整,浑厚朴实,尤其是砖上模印的各式花纹,更是以细密的纹理效果产生的丰富视觉层次感著称于世。成语"秦砖汉瓦"即是对秦代烧制的砖和汉代生产的瓦的赞誉。

西汉时,我国出现了装饰性较强的画像砖。画像砖是指在墓室中用于贴墙或加固的面砖,因其砖面往往有模印或刻有画像和花纹,故而被称为画像砖。西汉画像砖基采用流水式模印的生产方式,使同一种内容和形象重复出现。东汉的画像砖用一块砖表现出一个完整而独立的画面,是真正意义上的单个独立艺术品,蕴含着更加丰富和深入的艺术意境。汉代画像砖是墓室预制构件的大型空心砖,生产方法是在湿的泥坯上用印模捺印各种图像。北宋时发展为砖雕,并成为墓室壁面的装饰品。在河南、山西、甘肃等地发掘的北宋墓室,三面墙壁均以砖雕贴砌而成。墓室内的砖雕数量、质量以及内容题材都取决于墓室主人的社会地位。常见的题材有墓室主人夫妇对坐,身旁有男仆托盘、侍女执壶等,再现了墓室主人生前的生活情景。金代墓室砖雕的内容更加丰富,技艺也有所提高。

魏晋以后,墓室砖雕仍在发展。金元时期,山西等地的墓室砖雕已发展到相当规模。建于大安二年(1210)的山西侯马董玘坚墓室,在墓室内壁不足4.7m^2的面积上布满砖雕,雕刻有仿木结构的斗拱、拱眼、藻井、大门、隔扇等,以及屏风、几凳、花卉、鸟禽、人物、演戏场面等图案。砖雕中对站立在戏台口的生、旦、净、末、丑等角色的演员的雕刻运用了圆雕技法,使人物形象栩栩如生。到了元代,墓室砖雕逐渐衰落。山西是我国戏曲发展的胜地,在墓葬中发现众多以戏曲为题材的砖雕。至明代,砖作为民居建筑材料被广泛应用于民间和宫廷,戏曲题材的砖雕也由墓室壁饰转变为普通建筑装饰。在各地出现的书院、会馆、宗祠、寺庙、戏院等建筑中,无论是官式建筑还是民居,砖雕都是它们主要的装饰形式之一。

清代的砖雕艺术发展到巅峰时期。随着商业贸易的流通,地域之间的文化也在进行着传播和融合。最早从事盐业活动和药材生意的山西、陕西商人把山西一带流行的砖雕艺术带到江淮流域,运用于会馆装饰中。比如,现苏州市戏曲博物馆的旧址就是三晋会馆,在其前门门楼、门额上就有戏曲人物的砖雕;甘肃天水市陕西会馆的门楼上也有同样的雕饰;原为山陕会馆的安徽亳州大关帝庙内的花戏楼,也因雕有戏曲故事而得名。北京紫禁城内墙面夹柱的通气孔也都镂雕着花鸟图案,牢固而美观。慈禧太后陵寝隆恩殿及其东西配殿的墙面也用砖雕贴砌而成。

在装饰砖的选材上,砖雕所用的砖坯泥土较细,先进行细筛、淘洗、精漂等工序,再经过多次加工,制成精制的砖坯;再后控制好火候,经过一定的时间,将砖坯烧制成为质地细腻、大小规整的青砖;然后用清水清洗、磨洗后,经过精雕细刻,根据不同的题材采用不同的雕刻手法,变无生命的泥土为精致神奇的砖雕艺术珍品。青砖在选料、成型、烧成等工序上要求严格,所以成型的青砖坚实而细腻,适宜雕刻。在艺术形式上,砖雕远近均适宜观赏,具有立体效果。在题材上,砖雕以龙凤呈祥、六合同春、三阳开泰、郭子仪上寿、麒麟送子、狮子滚绣球、松柏、兰花、竹、山茶、菊花、荷花、鲤鱼等寓意吉祥和人们喜闻乐见的内容为主。砖雕饰制作的技法主要有阴刻(刻划轮廓,如同绘画中的勾勒)、压地隐起的浅浮雕、深浮雕、圆雕、镂雕(透雕)、平面雕(又称减地平雕,阴线刻划形象轮廓,并在形象轮廓以外的空地凿低铲平)等。民间砖雕从实用和观赏的角度出发,形象简练,风格浑厚,不盲目追求精巧和纤细,以保持建筑构件的坚固,能经受日晒雨淋。

(二)南徽北晋的砖雕风格

我国砖雕艺术不仅历史悠久,而且融合了南北方的地域和民族文化。不同地域的砖雕艺术风格既有融合之处,又显示出各自的特征。我国砖雕普遍得到认可的主要流派有:(1)北京砖雕;(2)天津砖雕;(3)山西砖雕;(4)徽州砖雕;(5)苏州砖雕;(6)广东砖雕;(7)临夏砖雕(又称河州砖雕,河州即甘肃河州地区;临夏,甘肃的古称)。在众多砖雕流派中,一般认为在中国南方发展最完善、成就最高的当数徽派砖雕以及受其影响演化发展而成的扬州、苏杭一带的江南民居砖雕。北方砖雕以山西、北京、天津砖雕为主，与南方砖雕并称为南徽北晋两大派。山西砖雕主要是晋商等富人所有的建筑内的砖雕，北京砖雕则主要指皇宫、皇家园林和达官贵人宅邸等建筑上的砖雕。晋派砖雕的特点是恢宏大气，刀刻线条粗犷明晰，构图浑朴厚重，内容相对简明独立；花卉图案繁复，多有凸雕的花瓣出现，而且墙字、匾额、楹联、碑刻较多；而徽派砖雕则刀工细腻精致，线条柔和纤细，花卉图案多层重叠，人物图案注重故事情节的完整和艺术语言的叙事功能。"总体而言，南方以苏州、徽州砖雕为代表，风格清秀雅致；北方以山西砖雕为代表，风格浑厚朴实。就艺术风格而言，北方浑厚，南方秀丽。①"

1. 徽派砖雕

徽州砖雕，即安徽徽州（今安徽歙县）的砖雕。"徽州砖雕源于宋代。明代尚古朴，以浮雕、单层次为主。清代趋向多层次透雕，技艺精细，用料多为水磨青砖"②，至明清而极盛。"明代雕刻粗犷古朴，一般只有平雕和浅浮雕，借助于线条造型，缺乏透视变化，但强调对称，富于装饰趣味。清代雕刻趋于细腻繁复，构图布局吸收了新安画派的表现手法，讲究艺术美，多用深浮雕和圆雕，注重镂空效果，有的镂空层次多达十余层，将亭台楼宇、山水树木、人物走兽、花鸟虫鱼在不同透雕层次上集中在同一画面上"③，犹如一幅浑然天成的水墨画，因具有独特的美感而名扬中外。亳州花戏楼砖雕中的《郭子仪上寿》是徽州砖雕的代表作，其高超的雕工令人惊叹。徽州地区的砖雕繁琐华丽，工艺精细，雕刻工整，运线流畅，主题突出，层次分明。因徽州建筑多用青顶白墙、青砖门罩、门楼和飞檐，砖雕装嵌其中，与徽州树木繁茂的山区自然环境十分协调。徽派砖雕的风格"注重情节和构图，往往以多层透雕为主。在尺余见方、厚不及寸的砖坯上雕出情节复杂、多层镂空的画面，前后层次分明，一块方砖坯上最多的可以透雕十几个层次。整个画面也采用了立轴和手卷式的布局，显得更加端庄严谨，令人产生精妙无比的美感，其技艺已到了'天工人可代，人工天不如'的艺术意境"④。

"门罩迷藻悦，照壁变雕墙"是徽州砖雕的应用的真实写照。徽州砖雕装饰的重点是门楼和门罩。作为古民居出入口标志的门楼、门罩造型多样。徽州民

---

① 毛晓青，王彩霞. 中国传统砖雕[M]. 北京：人民美术出版社，2008. 29，77，30.

② 鲍义来. 徽州文化全书·徽州工艺[M]. 合肥：安徽人民出版社，2005. 156.

③ 百度. 立体的画，无声的诗——精巧绝伦的各地砖雕[J]. 科学之友. 2011(3)：19－21.

④ 詹学军. 徽州砖雕的源流与艺术特点[J]. 美术大观. 2012(4)：70－71.

宅门楼上的门罩即是在大门外框上方用水磨青砖砌成的向外凸出的线脚装饰，顶上附以瓦檐。如潘氏宗祠，门厅为五凤楼饰式，气势壮观，门厅两侧的八字墙上装饰有大面积精美细腻的砖雕，描以额枋和框，内容有江南风光、楼阁亭台、水榭、飞禽走兽等，运用透雕和丰圆雕手法，高低起伏有度，别具一格，犹如一幅水墨画，清新淡雅。砖雕门楼的额枋，尤其是"额枋通景图往往是最精彩的部分。额枋通景图通常由五至七块砖拼成，雕有历史人物、山水名胜、钟鼎博古、瓜果花卉等，景中有景，寓意深刻。[①]"额枋通景图是在一个平面上，向里深雕六七个层次，运用浅浮雕、高浮雕、透雕、半圆雕和镂空雕等手法，突出雕刻主题，使其具有距离感、层次感。另外，脊饰和墙上砖雕也很常见，在大量寺庙、祠堂、民宅等不同的建筑上可以看到各式屋顶的正脊和正吻(正吻为正脊两端的雕饰)砖雕。

徽州砖雕图内容很广泛，有花鸟、人物、生活场景和吉祥纹饰等。以人物为主的题材包括历史故事、戏曲、宗教神话、民间传说和其他社会生活等，描绘的人物有帝王、贵族、文人、商贾、书生、樵夫、农夫、牧童等等。以动物为题材的象征吉祥的图案在门楼、门罩雀替、角缘部位较多，如《龙凤呈祥》、《鹿鹤同寿》、《丹凤朝阳》等。花卉图案更是丰富多彩，"岁寒三友"、"花中四君子"等都是十分常见的雕刻对象。这些形象大都是用折枝、散花、丛花、锦地叠花、二方连续、四方连续等手法，寓意喜庆、幸福，表现人们的美好愿望，达到了"图面有意，意在吉祥"的境界。受徽州砖雕影响而发展起来的江南地区的砖雕，造型更显雅致灵巧，清新细腻，刀法更加工整。江南民居砖雕风格纤细、刻工精良、空间层次丰富、意境深远，富于文人趣味。

岭南一带民居的砖雕手法更自由，体裁更丰富，民俗趣味更浓厚。岭南砖雕尤以广州陈家祠堂面对来客的三块大型砖雕为典型代表，东墙上的《刘庆伏狼狗》刻有30多个人物，西墙上《梁山泊好汉》砖雕作刻有十多个人物，他们姿态各异，服装不同，表情有别，都被刻画得清清楚楚，细致入微。两幅画采取浮雕、透雕、立雕等多种技法表现人物与环境的关系，使他们共处于多层次的空间中，井然有序。"较之北方砖雕的粗犷、浑厚，广东砖雕显出纤巧、玲珑的特点，往往雕镂得精细如丝，被称为'挂线砖雕'。雕刻手法多以阴刻、浅浮雕、高浮雕、透雕穿插进行，精细者可达7~8层，达到表现景致深远的效果。雕成的花卉枝叶繁

① 毛晓青，王彩霞. 中国传统砖雕[M]. 北京：人民美术出版社，2008. 29，77，30.

茂,形如锦绣,戏曲人物衣甲清晰。在不同时辰的日光照射之下,还能呈现出黑、白、青灰等不同色泽,高光部更袅袅生辉,画面富于起伏变化。[①]”

2. 晋派砖雕

晋中、京津等地的砖雕工艺纯熟,造型简练,风格质朴,庄重浑厚。晋派砖雕讲究构图大气庄重,刻形精美浑圆,密而有形,纹饰繁缛,刀法浑厚朴茂,雄浑之余透出粗犷之气。它的特点是砖雕坯胎的土质好,经久耐用;砖雕花样繁多,布局大气,画工精细,刀工别致,而且谱系明确,传承有序。著名的砖雕有山西常家大院、乔家大院和北京故宫的栏杆砖雕,颐和园的墙上砖雕、砖雕气孔、影壁等传世艺术作品。晋中、京津砖雕中的墙上砖雕、栏杆砖雕和影壁砖雕比在中国南方更为常见,且雕刻技艺与徽雕相比,具有拙朴、苍厚、稳健的砖雕艺术风格。

在中国北方,砖雕影壁是常见的建筑装饰。在北方,组合型建筑四合院是最常见的建筑范式,体现了中国传统建筑的布局特色。“中国地处北半球中纬度和低纬度地区,这种自然地理环境决定房屋朝南可以冬季背风朝阳,夏季迎风纳凉。所以中国房屋基本以南向为主,坐北朝南是中国传统风水理论的建筑原则之一”[②]。但是受地理位置和自然环境的影响和限制,中国南北建筑在面积、规模和房屋的朝向上呈现出一些差异。通常南方的建筑,祠堂、寺庙、民居等依山而建,因建筑面积受限,建筑布局密集紧凑,故常见之精巧雅致,婉转曲折,式样繁复。北方平原多、地域辽阔,常见的建筑布局是坐北朝南的宽敞的四合院。相比南方,北方场地宽敞,皇室官府宅邸众多,使用砖雕影壁作为建筑装饰很常见。

影壁是一面单体墙体,独立于房屋之外,位置在四合院或一组建筑的大门外或门内,隔一段距离正对大门。大门外的影壁多出现在官府、寺庙、大宅第门外,有标明大门位置的作用,使过往行人避开。大门内的影壁的功能是遮挡人们的视线,不让人一眼望到院内,从而保持建筑内部的隐蔽与安静。根据内外影壁的这一功能,原来分别称它们为“隐”与“避”,合称为“隐蔽”,后来演变为“影壁”。影壁不论位于大门外或门内,都是与进出大门的人打照面的,所以又称照壁。砖雕影壁既有“隐蔽”院内人的活动及景物的功能,又具有装饰性,还可以增强大门的气势,显示主人生活殷实,富有追求。因此,在北方,上至皇宫皇室,下到市井民居,砖雕影壁成为了一种重要的建筑装饰。如,山西乔家大院有一座雕刻着

① 百度. 立体的画,无声的诗——精巧绝伦的各地砖雕[J]. 科学之友. 2011(3):19-21.

② 魏宪田,黎光. 相宅者说[M]. 杭州:中国物质出版社,2010. 49.

一百个篆体寿字的影壁,横竖各十个,黑底金字,字体各异,显得庄严厚重。在众多的砖雕影壁中,设计最为大气恢宏的当属山西常家大院的砖雕影壁。大院横向很宽,共有大小六处影壁。大院内院一处有两座垂花门并列,两门中间有一座一主两从的大影壁,两门的外侧各有一个影壁,共有一大四小五座影壁。影壁古朴大方,浑厚苍健。每个影壁的壁座部分都由雕刻精美的须弥座衬托,犹如五幅雕花的座屏,气势恢宏,令人赞叹。

(三)花戏楼的砖雕艺术

中国古代建筑数千年的发展历史为后人留下了无数辉煌的宫殿、宏伟的寺庙、秀丽的园林和千姿百态的住宅。在这些建筑中,装饰无疑起到了十分重要的作用。就戏楼通体的砖雕装饰艺术而言,亳州花戏楼三层牌坊式建筑的山门上镶嵌的立体水磨砖雕不仅雕刻得精美绝伦,而且富含耐人寻味的深奥文化内容。"砖木二雕为该庙之精华所在,为研究清中叶建筑艺术之重要实例[①]"。青灰色的砖雕,玲珑剔透,精美秀丽,透出一种朴素无华的厚重美感。砖雕的伦理文化主题突出,艺术特色鲜明。其表现之一就是表达颂扬伦理道德内容的戏文、故事和人物作品被装饰在建筑主体部位,旨在宣扬忠孝礼义等传统道德。这些作品在人物周围添加了背景,如花卉草木、桥梁、房舍、庭院、池塘等,或配有珍奇动物和神兽,烘托装饰环境,构成十分和谐自然的画面。整幅砖雕组图内容丰富,构图朴实,布局巧妙,比例适当,结构对称,且刀法细腻、雕工精湛,堪称传世杰作,在全国范围内的会馆砖雕装饰艺术中具有非常重要的地位。

凡是到过花戏楼的人,亲眼目睹那刻满故事的精美砖雕,欣赏着那方寸之地展现在我们面前的大千世界,无不感叹于那精美砖雕装饰给人带来的艺术震撼力。遗憾的是,据安徽省旅游局文物管理专家考证,亳州花戏楼水磨青砖砖雕工艺技术目前已经失传。因此,现存的亳州花戏楼砖雕作品已经成为国内乃至世界范围内的艺术绝唱。

"砖雕的定义就是以砖为原料雕琢的建筑构件。砖雕一般用在屋顶的瓦作、塔身的装饰、园林中的门楼和漏窗、大院里的照壁等。所以说到砖雕就不能脱离建筑物本身的特征。[②]"砖雕用于装饰墙面时,按装饰的部位分为檐墙砖雕、

① 陈从周. 亳州大关帝庙[J]. 同济大学学报. 1980(2),85.
② 毛晓青,王彩霞. 中国传统砖雕[M]. 北京:人民美术出版社,2008. 29,77,30.

廊心墙砖雕、山墙砖雕、院墙砖雕等类型。亳州花戏楼大门上三层牌坊式的青砖雕刻作品属于山墙砖雕。花戏楼所在的建筑群（即山陕会馆）虽然兼具有商业、娱乐、政治等方面的功能，但其主要作用是祭祖，故花戏楼的大门常被称为山门。山门原意为寺院正面的楼门，它是对寺院大门的一般称呼。过去的寺院，通常为了避开市井尘俗而建于山林，因此称山号、设山门。花戏楼虽造于亳州古城的平地，但因其内建有祭祀关羽的大关帝庙，大门亦被称为山门。在宗教意义上，山门等同于“三门”，象征佛教的“三解脱门”，即“空门”、“无相门”和“无作门”。今花戏楼山门就由中间的正门、东西两侧的钟楼门和鼓楼门组成。精美的砖雕组图位于山门上方，山门背后是花团锦簇的戏楼。

花戏楼山门楼牌坊整个面积仅有十平方米，在厚度不足两寸的水磨青砖上共雕有52幅作品，雕刻人物115个，禽鸟33只，走兽67只，此外，还有种类繁多的山水桥洞、林木花卉、车马城池、亭台楼阁。砖雕的内容十分丰富，有戏文6出，人物故事16幅，动物典故24幅。装饰图案如纹饰、织锦、葵花、祥云、缠枝花卉，福寿字等，锦上添花。山门正面一块名为《郭子仪上寿》的砖雕最为显眼。郭子仪是唐朝的一位名将，因平定安史之乱有功，被唐玄宗封为汾阳王。图案雕刻的是郭子仪六十大寿时文武百官、七子八婿为他祝寿的喜庆场面。整幅砖雕作品刻有42个人物。人物形态各异，但都满面春风，喜气洋洋。有的扶老携幼，有的坐轿骑马。画中的郭子仪端坐在正堂，胡须垂胸，慈眉善目，面带微笑，和蔼可亲。他身后有一个大“寿”字清晰可见。朝中文武官员依次站立，或手捧贡品，或俯身作揖，每个人的表情各具特色，耐人寻味。砖雕画面两侧亭台楼阁玉立，车马人流熙攘，显示出一派富足祥和的景象。整个雕刻布局均匀，比例适宜，刻画细致入微，令人称绝。

花戏楼山门上的砖雕牌坊在方寸之地展现出大千世界的奇妙故事。正门砖雕序列左右两侧有反映中国文人生活情趣的“四爱图”:《王羲之爱鹅》、《周敦颐爱莲》、《鲁隐公观鱼》和《陶渊明爱菊》。《虎落平阳》出自于古代成语故事“虎落平阳”。从砖雕上看到老虎站在平原上，威风扫地，而两只狗却精神抖擞地对失势的老虎狂吠。砖雕的寓意深刻，比喻人在失势后遭到冷遇的处境，揭露了一些人趋炎附势的本性和世态炎凉。这句话出自中国古代儿童启蒙书《增广贤文》，最早见之于明代万历年间的戏曲《牡丹亭》，后经明清两代文人的增补，成为极具智慧和哲理的民间创作读本。九狮图和五狮图中的“狮”的谐音为“世”，《九世同居》，即九代同堂;《五世其昌》，意味着人丁旺盛，世代昌盛，还有事事如意之含义。山门正面雕刻的还有《魁星点状元》、《衔环报恩》、《六合同春》、《心猿意马》、《鹿骇狼顾》、《万象更新》、《犀牛望月》、《怒蟾斗狮》等,都出自于中国典故或成语,寓意深刻,耐人寻味。

鼓楼上面是一幅令人叹为观止的《三顾茅庐》雕刻作品,讲的是三国时期刘备三请诸葛亮的故事。砖雕上左边三个人物即是刘备、关羽、张飞,迎面背手而去的是诸葛亮的弟弟诸葛均。历史故事中有如下描述。刘备上前问话:"令兄可在? 可得一见?"诸葛均曰:"吾兄昨晚方归,先生可见。"说罢背手而去,态度轻慢。图中右上方在睡榻上睡觉的便是诸葛孔明。"大梦谁先觉,平生我自知,草堂春睡足,窗外日迟迟。"这首诗就是孔明当时吟诵的。这幅砖雕雕工手法细腻,层次分明,艺术价值极高。从画面上可以看到一茶童在榻前煮茶,旁边有一个水缸、两只水桶。煮茶的茶壶、书斋中的桌椅和笔墨书籍都清晰可见。榻前右上方有透刻的窗棱和雨搭,细如火柴杆的竹竿支撑着雨搭,远近分明。诸葛亮的一双鞋子摆在床榻前,非常逼真。

"白蛇传"的故事在中国家喻户晓,而钟楼下方的《白蛇传》砖雕上的人物形象更是生动逼真,景物刻画也是十分精细,具有通透的立体感和真实感。你可以看见断桥边的流水和雷峰塔,庙门敞开的金山寺。想象通过庙门拾级而上进入庙院,如同身临其境,引发无限遐想和感慨。

亳州花戏楼的砖雕装饰大多采用组合型砖雕,即由多块青砖雕刻构成的一个完整的砖雕构件。雕刻手法有浅浮雕、高浮雕、透雕、圆雕、镂空雕和线刻等。雕饰技艺为体现材质美和工艺美,运用对称、呼应、疏密、虚实、明暗、刚柔等对比因素,同时考虑布局、位置、比例、大小等关系,力求主次分明,强调立体感、空间感、韵律感等形式美感。花戏楼整座砖雕牌坊雕刻技艺精湛,工艺流程复杂,刀法细腻,所塑形象生动,雕刻线条变化有度、流畅自然,场景布局安排错落有致,是清代砖雕的典型代表,"具有很强的艺术感染力和视觉冲击力,在国内很罕见"①。亳州花戏楼砖雕牌坊也是大关帝庙的精华所在,"从风格上看,融合了徽派砖雕的精细和晋派砖雕的浑厚"②。水磨砖雕吸收了两派雕刻的艺术特点,产

① CCTV 4. 国宝档案 · 亳州寻珍——清代花戏楼砖雕. 解说词[OL]. 2013 - 6 - 9. The Gorgeous Dramatic Stage.

② CCTV 4. 国宝档案 · 亳州寻珍——清代花戏楼砖雕. 解说词[OL]. 2013 - 6 - 9. The Gorgeous Dramatic Stage.

生了一种刚柔相济的工艺之美。花戏楼砖雕的另一个艺术特色就是形式与内容的协调。砖雕画面结构完整,文化意蕴丰富。砖雕这门高雅艺术走出艺术殿堂,与市井文化相结合,寓教于乐,集教化、娱乐、宣扬、崇拜和告诫功能为一体,达到了艺术审美最高境界,起到了文化熏染作用。

亳州花戏楼在没有任何保护的情况下历经了三百多年的寒来暑往、日晒雨淋,它的石狮和铁旗均有部分被风化,唯此砖雕安然无恙,完好无损,可见我国古代烧砖技术之高超,令人叹为观止。据说其奥秘在于砖雕的雕刻材料。砖雕与木雕相比有两个优点,一是材质防潮防火易保存,二是材质柔软易雕刻。亳州花戏楼砖雕使用的青砖在烧制过程中,"掺进了发丝、棉絮等调和物。烧制成的青砖具有很强的抗腐蚀能力。正是有了这种特殊的工艺,这巧夺天工的砖雕才得以保存至今。①"花戏楼砖雕不仅是艺术珍品,更凝聚了中华文化之精华,正可谓岁月无痕,精华留存。如今,砖雕工艺已经失传,花戏楼精美的砖雕已成为世间珍品,无法加以复制。

三、花戏楼建造始末

随着现代化的进程,城市的发展和人类的活动正在影响着中华大地上所有的古代建筑和其他文化古迹。当昔日的辉煌和繁华散去,依然巍峨地坐落在亳州城北关的花戏楼,其精美的砖雕木刻、繁花似锦的戏台和鲜艳的雕梁画柱魅力依旧,风韵犹存。这座存在了300多年的戏楼与芍药、华佗及亳州的中草药栽培历史有着不解之缘。

(一)"药材之乡"的由来

亳州中药材的发展和经营历史悠久。自东汉以来,亳州(谯县)就有人从事中药材的种植、炮制和经营,亳州也被称为中国古代的"芍药之乡"。明清时代,随着牡丹栽培中心的移入,亳州很快发展成为全国闻名的中药材集散地,奠定了它"药都"的地位。这与亳州人从古至今种植芍药、牡丹的历史有着密切关系。

古时候,芍药与牡丹没有被清楚区分。芍药、牡丹均为芍药科植物,两者花形相似,都有香味。前者为草本植物,后者为木本植物。芍药品种繁多,浑身是

---

①　CCTV 4. 国宝档案·亳州寻珍——清代花戏楼砖雕. 解说词[OL]. 2013－6－9. The Gorgeous Dramatic Stage.

宝。芍药花盘大而艳丽,在园林中常成片种植,花开时十分壮观。芍药花也可用来切花插瓶,极具观赏性。芍药也是观叶植物,诗句"红灯烁烁绿盘龙"中"绿盘龙",就是对芍药叶的赞美。芍药的根茎可制成中药,具有镇痉、镇痛的作用。因此,芍药被誉为"花仙"和"花相",还被列为中国"六大名花"之一。据《本草》记载:"芍药犹绰约也,美好貌。此草花容绰约,故以为名。"

芍药是中国栽培最早的一种花卉。在秦以前的典籍中,只有芍药而没有牡丹的记载。公元前770－221年的春秋战国时期,我国第一部诗歌总集《诗经·郑风·溱洧》有诗句曰:"维士与女,伊其相谑,赠之以芍药。""据此推知,2500多年前的中国古代,芍药花就被当作男女交往的爱情信物,表达结情之约或惜别之情"[①],故芍药花又称"将离草"。因其自古就被认为是爱情之花,现已被尊为七夕节的代表花卉。

明人鸿胪寺少卿薛凤翔所著的《牡丹史》记载:"芍药著于三代之际,风雅所流咏也。今人贵牡丹而贱芍药,不知牡丹初无名,以花相类,故依芍药为名。"《古琴疏》载:"夏帝相元年,条谷贡桐芍药。帝命羿植桐于云和,命武罗伯植芍药于后苑。"至今我国的芍药栽培历史超过4000年。亳州的芍药栽培历史可追溯到西周。唐代《通典》记载:"周初,武王克殷,封神农氏之后于焦。"焦,即亳州。"西周初期,神农氏后裔姜氏即在亳州城东北隅修神农衣冠冢,建先医庙,并教人种植草药以治疗疾病。"《亳州志》记载,公元200年,亳州境内开始广植芍药等中药材,开启了亳州芍药和其他药草的种植历史。三国时期,曹操之子、魏文帝曹丕(亳州人)在《皇览》一书中也曾对芍药作过专门论述。公元前2世纪,秦灭楚,设谯县,亳州遂有"谯"之称。东晋时期成帝在谯县留郡,改谯县为小黄县,从此谯县又称小黄县。清代诗人刘开路过亳州时,看到家家户户的房前屋后种植白芍,赋诗曰:"小黄城外芍药花,十里五里生朝霞,花前花后皆人家,家家种花如桑麻?"可见当时亳州的芍药生产种植已经很兴旺,已成为远近闻名的"芍药之乡"。

牡丹的出现是在东汉时期。1972年甘肃省武威市出土的东汉早期圹墓中的医简载有牡丹治疗"血淤病"的处方。这是迄今有关牡丹的最早记载。古人在民间引植、栽培原始牡丹大概始于南北朝。史书《牡丹赋》中记载"天后之乡,

---

① 王正明,魏宏灿,张立驰.亳文化十讲[M].合肥:安徽教育出版社,2013.137－139.

西河也,精舍下有牡丹,其花特异,天后上苑只有缺,因命人移栽焉。”到南北朝时,牡丹已成为观赏植物,大约在隋唐时进入宫苑。据《隋炀帝海山记·炀帝宫中花木》记述,当时,“帝辟二百里为西苑,诏天下鸟兽草木”,易州(今河北易县)进牡丹二十种,隋炀帝还一度称牡丹为隋朝花。北宋年间,洛阳种植牡丹规模空前,栽培技术也更加系统、完善,培养出许多名贵品种。牡丹被称为洛阳花、京花。到唐代开元中,牡丹已盛极一时,其栽培中心从长安东移到洛阳。诗人刘禹锡在《赏牡丹诗》中谓:“唯有牡丹真国色,花开时节动京城。”明朝(1368~1644年)牡丹栽培中心从都城洛阳转移到安徽亳州。史书记载:“洛阳牡丹始于隋,盛于唐,甲天下于宋。”

亳州的牡丹栽培历史及牡丹的兴盛与我国明代杰出的牡丹园艺学家薛凤翔及其家庭有着密切联系。薛凤翔,字公仪,安徽亳州人,出自于喜爱牡丹的世家,约生活在明代万历年间,官至五品鸿胪寺少卿。其祖父薛蕙(1489~1541年)祖居亳州城内薛家巷,曾任明代吏部考功司郎中,对其牡丹园艺的爱好影响极大。“公自吏部归,绝意仕进,营园自适”,置有《常乐园》;其父号两泉,置有《南园》。薛凤翔出身仕宦望族,家有名园,广植牡丹。他受家庭影响,在绩学之余以种花自娱,对牡丹史料和栽种之术了如指掌。“栽花万万本,而牡丹为最盛”。他辑录了历代有关牡丹的资料,总结亳州花农的栽培技术和经验成果,加上自己的研究成果和心得,“穷其变态著而为史,是为《牡丹史》”。《亳州牡丹史》则为《牡丹史》卷一中的传六部分,后人又称《牡丹史》为《亳州牡丹史》。《亳州牡丹史》内容丰富,富有文采,兼文学性、科学性、艺术性于一身。该书详尽记述了明代亳州牡丹的兴盛历史,成为有史以来我国研究牡丹的最著名的一部专著。它从牡丹品种、栽培技术、亳州牡丹的奇闻掌故等方面,证实了在明代我国牡丹的栽培和研究中心已转移到亳州。

薛凤翔在其所著的《牡丹史》本纪中写道:“独怪永叔尝知亳州,记中无一言及之,岂当时亳无牡丹耶。德靖间,余先大父西原、东郊二公最嗜此花,偏(遍)求他郡善本移亳中,亳有牡丹,自此始。顾其名品仅能得欧之半。”永叔即欧阳修。德靖间,指正德(1506~1521年)至嘉靖(1522~1566年)年间。二公,指西原,即薛凤翔的爷爷,东郊即薛蕙的弟弟。薛蕙还乡闲居后,与其弟(号东郊)在亳县城南营筑“独乐园”,后改名为“常乐园”。园中广植牡丹,极尽名种,名盛一时。“在薛氏祖孙三代的影响下,亳州牡丹园艺业蓬勃兴起,以至于到薛凤翔时,亳地牡丹名噪海内"[①],"每至春暮,名园古刹,灿然若锦,可谓盛况空前。[②]"《牡丹史》记录"亳以牡丹相尚,实百恒情","计一岁中,鲜不以花为事者"。风俗记中说,"又截大竹贮水,折花之冠绝者斗丽",如痴若狂。亳州的牡丹名园有十四所之多,各具特色。花师辈出,"种艺竟巧,相尚成风"。亳州人爱好牡丹已蔚然成风,种植培育、研究赏鉴者众多。

亳州芍药和牡丹栽培的兴盛,引起了亳州作为当时中药材交易地的兴起和繁荣,而东汉神医华佗的成就也得益于"药材之乡"得天独厚的自然与人文环境。华佗(约145~208年),东汉末年著名医学家,字元化,一名旉。沛国谯(今安徽亳州)人。华佗与董奉、张仲景(张机)并称为"建安三神医"。华佗不求仕途,在自家房前屋后开辟药圃、药池种植草药,自己开设医馆,专志于岐黄(中医)之术,行医足迹遍及安徽、河南、山东、江苏等地,被后人称为"外科鼻祖"和"华佗老祖"。华佗对我国中医学的贡献主要有三方面,"一,发明并在手术中使用'麻沸散',开创了世界医学史上最早应用全身麻醉进行手术治病的先例。二,根据'五行学说'和中医理论,模仿虎、鹿、熊、猿、鸟等禽兽的姿态,创作了名为'五禽戏'的健身操,开创了体育健身的先河。'五禽戏'融气功和武术于一体,有舒筋、活络、畅气、壮骨之效,可以强健体魄。三,依据农时农事规律,引进野生草药进行人工栽培,提高药草的质量,用于治病防病。[③]"亳州土地肥沃,四季分明,雨水充沛,非常适合中草药生长,种植中草药逐渐成为亳州人从事的一项主要农事,经营中药材之风也日趋渐盛,亳州自然地成为了药材集散地,至今仍然如此。

如今,亳州成为全国闻名的四大药都之一。亳州栽培的白芍始于魏晋,现已成为全国闻名的地道药材。亳州白芍以年产量占全国70%的比例位于全国白芍产量之首。芍药花成为亳州的市花,人们为它铸成一座高20余米的"芍花王"不锈钢雕塑,矗立于亳州市魏武大道。前总书记江泽民亲自为亳州题写了"华佗故里,药材之乡"的题词。亳州市每年九月九日举行的亳州中药材交易会更是吸引了来自于海内外各地的中药材商人来亳州洽谈生意、切磋药艺、畅叙友情。

---

① 潘法连. 薛凤翔及其《牡丹史》[J]. 中国农史. 1986(4). 45-46.

② 吴诗华. 薛凤翔与《亳州牡丹史》[J]. 中国园林第7卷(2). 54.

③ 亳州市文联,亳州市旅游局. 亳州之旅[M]. 北京:中国文化出版社,2009. 48.

(二)花戏楼(山陕会馆)的修建

花戏楼是亳州山陕会馆,大关帝庙(1656年)建筑的一部分,其增建时间在会馆以后(1756年)。会馆类建筑象征着资本主义萌芽,是商业繁盛的产物。会馆即聚会之馆,始建于明代前期,大约在15世纪的明代中叶才较多出现,到清代特别盛行。明清时期,经历了金、元、宋各朝代的文化传播和融合,中国的经济和文化开始繁荣,南北经济往来、文化交流增多。明代中叶以后,城市商业经济进一步发展,资本主义萌芽已经出现,各地商人聚集于城市或交通要地。为了在商业竞争中获取更多信息,一些同乡或同行业的商人常举办聚会交流信息。会馆应运而生,并且很快发展起来。"从实际用途来看,会馆分为两种:一种是政客们建的"同乡会馆",同政治、文化关系密切。一种是商贾们修建的"同业会馆",与商业、经济联系密切"①。同乡会馆提供膳食住宿,接待前来会考、游历、读书的同乡,供同乡人聚会。同业会馆则以行业名称命名,接待同行或用于聚会。也有两种性质兼有的会馆,接待同乡的同行商人。分散于中国各地的山陕会馆就是两种功能兼备,既有商务又有文化用途的会馆。

山西、陕西两省在明清时代形成的两大驰名天下的商帮——晋商与秦商,合称为"西商",或称"晋商"。山西和陕西仅一河之隔,自古就有秦晋结好的佳话。成语"秦晋之好"即缘于此。明清时期,山西与陕西商人常利用邻省之便,互相合作,组成商业联盟,以对抗徽商及其他商人。晋商是一个凝聚力很强的群体,他们非常重视与家族、乡亲及社会各方面的协调,相互之间既有竞争关系,更有相互照应,互相帮助的关系。由于陆路和水路交通提供了便利条件以及自然资源丰富等原因,中国南方的商品经济开始变得繁荣。山陕商人经常往返于山陕家乡和经济发达的南方,从南方贩运药材、食盐、茶叶等商品到北方。中国中东部平原、云贵高原的一些交通比较便利的城镇,便成了西商(晋商)的落脚地。商人们远离故土在异地经商,商场往往危机四伏,使人如履薄冰。为了让乡友们找到一种精神寄托和归宿,山陕商人们自筹资金,在当地修建会馆,设立同乡会、行会,并制定行规来处理商务纠纷,维护公平竞争。山陕会馆同时又是山陕商人"敬关公、安旅故、叙乡谊、通商情、立商规、兴义举"的场所,如河南的社旗山陕

---

① 程裕祯. 中国文化要略[M]. 北京:外语教学与研究出版社,2011. 264 - 265.

会馆,山东的聊城山陕会馆,湖北的襄阳会馆等,也有的叫山西会馆、西商会馆或秦晋会馆。会馆既可以作为联络乡谊,方便仕商,维护集团利益,聚会议事,交流信息,购置冢地,举办慈善活动和娱乐休息的场所,同时又是一个自主并自我约束的社团组织所在地。会馆使在外经商的晋商形成一股强大的凝聚力量,为西商在中国商界称雄 500 年的辉煌历史做出了贡献。

在五个多世纪里,山陕商人从盐业起步,逐步将经营范围扩大到棉、布、粮、油、茶、药材、皮毛、金融等各个行业,并且在每个行业都能雄踞一方。山陕商人将商业活动从故乡扩展到全国各地的重要城镇和商埠都会。他们从国内贸易开始做起,后来又把贸易扩展到蒙古、俄罗斯、朝鲜等邻近国家地区。山陕商人的贸易活动,大大丰富了中国古代的商业文化,把中国的贸易推向了一个新的高度。明清时期的山陕商人把山陕两地经商的智慧和艺术推向了极致,山陕商人的魄力之大、足迹之远、财富之巨让世人认同了“无西不成商”(“西”指山陕商人)的说法。

西商作为中国三大商帮之首,经济实力雄厚,富甲一方。在各地建造会馆时,客居各地的富商大贾不惜投入重金为建筑会馆提供强大的财力支持。据有关碑文记载,号称“中国会馆之最”的南阳社旗山陕会馆,“第一期工程兴建春秋楼及其附属建筑,花费白银 707844 两;第二期工程兴建大拜殿及其附属建筑,花费白银 87788 两。[①]”在各地现存的山陕会馆中,人们都可欣赏到极其考究的三雕建筑装饰,其装饰技艺之精湛,手法之讲究,令人叹为观止。那些宏伟壮观、金碧辉煌的会馆,无不显示出西商雄厚的经济实力。

亳州的花戏楼又称山陕会馆,是众多的明清山陕会馆之一。作为商朝的都城,亳州自古就是一大商埠,商贸繁荣,文化多元,其“‘好尚稼穑,重于礼义,厚重商贾,机巧成俗’的亳州地方文化[②]”历代延续。明清时期,亳州中医药生产和中药材交易已全国闻名,成为我国中原最大的中药材集散地,南北各地药商纷纷来亳开店设坊,从事中医药经营。中药材交易所遍及亳州大街小巷,街上前店后坊的药店、药铺、药行,鳞次栉比,相互毗连,仅北关就有上百家中药材商号、店铺。亳州出产的药材有 170 多种。《药典》上以“亳”字命名的“亳菊”、“亳芍”、“亳桑皮”、“亳花粉”四种药材是医家推崇的上品。尤其是亳州地产的白芍,色

① 赵静. 中国赊店. 山陕会馆[M]. 郑州:中州古籍出版社,2013. 2-3.

② 任晓民. 亳州 · 名城 · 名胜[M]. 香港:香港天马图书有限公司,2002. 41,39,115.

白如玉，粉性足，疗效好，与杭州的“杭芍”和四川的“川芍”并列为中国三大名芍。亳州众多的药草品种和兴旺的药材生意吸引了全国各地的药商前来经营，其中自然少不了晋商。

古代的亳州地处中原，位于中国南北部和东西部的交汇处，地理位置优越。北有涡河上连黄河、下通淮河，水陆运输便捷，商业历史悠久。早在春秋战国时期，亳州就成为了楚、宋、鲁等国的商品集散地。谯县为谯郡治所时，商业得到了发展。“唐代亳州为天下‘十望’州府之一，北宋成为南北货物集散地，土产绉纱设有专柜买卖。到了明清，城内会馆、商店遍布，亳州商业发展到了兴盛时期，成为苏、鲁、豫、皖四省的物资集散地，被誉为‘小南京’[①]。”《亳州志》记载了当时的街市：“商贩土著者，什之三四，其余皆客户，北关以外，列肆而居，每一街为一物，大有货别队分之气象。关东西，山左右，江南北，百货汇于斯，亦分于斯。客户既集百货之精，目染耳濡，故居民之服食器用，亦杂五方之习。[②]”城内商品多集中经营，商行店铺云集，贸易空前繁荣，出现了白布大街、竹货街、打铜巷、帽铺街、干鱼市、牛市、驴市等商品专营市场和纸坊街、里仁街、老花市等近百家药栈。

《亳州志》对亳州古城曾经的繁华这样描述道：“豪商富贾，比屋而居，高舸大艑，连樯而集，时则锦幄为云，银灯不夜，游人之至者相与接席，摧觞徵歌啜茗。一喙一蹴一箸之需，无不价踊百倍，浃旬喧宴，岁以为常。”各地经营药材、皮革、干果、杂货、棉纺、绸缎等的商帮纷纷在亳建立会馆，最多时会馆数量达到30多座。在徽州会馆、江宁会馆、福建会馆、浙江会馆、山东会馆、河南会馆、金陵会馆等之外，还有粮坊会馆、染商会馆、皮厂会馆、杀猪会馆和南京巷钱庄等。建造规模最大、最富丽堂皇并且留存至今的是位于亳州古城北关，原名大关帝庙的亳州山陕会馆。如今，关帝庙的宗教意义已经变得淡薄，会馆的功能也不复存在，人们更习惯称它为“花戏楼”。

会馆由山西商人王璧、陕西商人朱孔领发起筹建。据记载，这两位商人在亳州做药材生意发家后，便萌生了建会馆的想法，最初建立的是关帝庙。其作用一是祭祀神灵；二是用于在亳经营药材的商务联络场所；三是为同乡们提供一个安身之所。亳州山陕会馆主殿大关帝庙始建于清顺治十三年（1656年）；清康熙十

---

① 任晓民.亳州·名城·名胜[M].香港：香港天马图书有限公司，2002.41，39，115.

② 杨传中.亳州花戏楼的音乐历史学价值[J].阜阳师范学院学报（社会科学版）.2011(3).31.

五年(1676年)庙内增建了繁华至今的戏楼;乾隆三十一年(1766年),戏楼增添了彩绘、雕刻。“藻彩歌台(戏楼),固已极规模之宏敞,金碧之辉煌矣。”《亳州志》记载,乾隆四十一年(1776年),“关帝庙特华,内及雕镂藻绘之工,游市廛者,每瞻观不能去。”据清乾隆三十八年(1773年)《重修大关帝庙记》碑文记载:“亳州北关之大关帝庙,建于国朝顺治十三年(1656年)”,“首事王璧、朱孔领二人皆系籍西陲,而行贾于亳,连袂偕来,指不胜屈,亟谋设会馆,以为盍簪之地”。修成之后,“壮丽恢宏,美哉伦奂矣”。首建之后,历经修缮,才形成今天的规模。有案可稽的修葺活动有:一修于康熙二年(1663年),二修于康熙二十三年(1694年),三修于康熙五十二年(1713年),四修于乾隆十九年(1754年),五修于乾隆三十一年(1766年),六修于乾隆四十一年(1776年)。之后在道光、光绪年间均有修葺,最后一次的彩绘是在光绪十八年(1892年)——彩绘中有落款可证。这些修缮活动跨越三个朝代,历经260年,均由山陕商人捐资完成。会馆历经多次扩建修缮,已走过351个年头。它见证了晋商在亳州的发展史。

明清时期是我国经济、文化发展的鼎盛时期,因此花戏楼的建筑格局和装饰风格处处显示出明清时期中国古代建筑恢弘大气的布局特点和精致典雅的装饰特点,也体现出封建等级制度下的建筑规范。在建筑装饰方面,花戏楼无论在色彩还是构图方面均属于殿堂式建筑,宏伟而华丽。但是,虽然戏楼的屋脊脊饰采用了黄色瓦,屋顶用的却是青瓦而不是黄色琉璃瓦。虽然花戏楼色彩华丽,装饰精美,但很少有五爪龙的雕饰出现。这也就体现出明清时代封建等级制度的森严。

在中国五千年文明的历史长河中,君主专制是国家的主要政体。“溥天之下,莫非王土,率土之滨,莫非王臣。”在这种政治体制下,君主赖以维护社会秩序的两样法宝是“礼”与“法”。《礼记》中称“礼为天下之序”。壁垒森严的封建专制体制下的中国传统建筑也被分为众多等级。从建筑的布局,大小,结构,材料到装饰色彩、图形的使用,处处体现着严格的规范。各朝统治者甚至以法律的形式对此加以规定。[①]”早在先秦时期,规范建筑等级的法令便已出现。就城市建筑而言,王与诸侯大夫等所居的城市建筑有严格的等级限制。《周礼》具体规定了城隅、宫隅的建制,如有违者,即视为越礼。例如,当郑国的段叔恃宠擅自扩

① 姜晓萍.中国传统建筑艺术[M].重庆:西南师范大学出版社,1998.10.

大城邑的规模时，就被郑国大夫祭申斥为僭礼越法。隋唐时期，封建典制日趋完善，统治者对民居的建筑风格也作出了明确规定。宋代对建筑等级的规定甚至包含了对建筑材料的限制。《营造法式》将建筑材料分为八等，规定只有殿阔为九至十二间的最高等级建筑可用一等材。如果大材小用或"小材大用"，皆为越礼之举。

明代有关建筑等级的限制更加严格，如《明会典》对不同级别的官员及庶民所盖房屋的间数、装饰颜色都有明确规定。清代将建筑分为三个等级：皇帝及其家属居住的地方为殿式建筑。这类建筑宏伟而华丽，可用黄琉璃瓦顶、斗拱、重檐、藻井，可绘以各式彩绘图案。各级官员与富商缙绅的居室为大式建筑，这类建筑虽然也装饰精美，但不许用黄色琉璃瓦，也不许描龙画凤。普通百姓的居室为小式建筑，这类建筑以实用为主，极少装饰，不许用斗拱、重檐等。此外，在建筑的门堂、开间、进深以及屋顶的式样、色彩、装饰等方面，清代统治者做了严格限制。富甲一方的晋商在建造山陕会馆时，也毫不例外地受到了这种约束。

亳州山陕会馆，大关帝庙，现名花戏楼，是依照中国传统建筑的特点进行布局的。讲究对称性和空间秩序美的建筑布局烘托出戏楼的宏伟气势。从建筑结构上来看，"花戏楼"和"大关帝庙"都不是对这组群体建筑的准确表述，因为"花戏楼"和"大关帝庙"只是亳州山陕会馆的两大主体建筑。大关帝庙修建在前，花戏楼补建在后。和其他地区的山陕会馆一样，大关帝庙显示出晋派建筑大气恢宏的风格，但在装饰艺术和雕刻技艺方面兼有徽派建筑细腻精致的风格。整个建筑以关帝庙大殿为主，戏楼与山门砖雕牌坊为辅，坐北朝南，属于典型的中国传统寺庙建筑。大殿坐北朝南，与戏楼相望，戏楼（歌台）坐南朝北与山门相背。坐北朝南的山门砖雕牌坊下有三个拱形门洞一字排列，中间门头上有刻着"大关帝庙"四个金字的扁额。山门左边是钟楼，右边是鼓楼。从山门砖雕牌坊下的拱门门洞进去，可以看到一条砖墁通道连接戏楼和大殿，它是建筑的中轴线。"轴线两翼分列钟楼、鼓楼、耳房、厢房、看楼，构成一个古老的四合院。院内南北屋顶起伏，东西看楼排列，廊庑周接，四环四合，井然有序，构成了高墙深院宁静舒适的环境。①"每当歌台演剧之时，乐声、鼓声、歌声响起，院内音乐环绕回响，音响效果良好。在科学技术不发达的古代，人们能设计出这样的建筑布

① 任晓民.亳州·名城·名胜[M].香港：香港天马图书有限公司，2002.41，39，115.

局,让现代人不禁感叹他们的聪明智慧。

花戏楼(山陕会馆)是晋商的历史和文化存在的非常重要的物证。它留给后人的不仅是美轮美奂的建筑艺术,更是一笔不可复制的文化遗产。会馆的建筑格局和关帝庙的装饰艺术,对研究明清时代的社会、经济、文化,以及晋商恪守的传统价值和经商理念都有着很重要的价值和意义。尤其是在花戏楼砖雕中,古代工匠们用他们精湛的技艺呈现出了煊赫一时的晋商的财富与人生理念,用象征性手法形象地传达出了晋商所奉行的儒学思想,为后人留下了久远的文化记忆。

第二章　花戏楼砖雕文化审美蕴涵

建筑被誉为“石头的史书”和“世界年鉴”，这表明它不仅是一门综合艺术，而且体现着人类的历史和文化。“中国传统建筑正是以其独特的语言方式，向人们倾诉着中华五千年的文明历程和炎黄子孙的思想情感。[①]”建筑凝聚着人类的创造和智慧，传递着人类的期盼与诉求，建筑文化是特定时期人们的社会生活和价值观的体现。从中国原始居民雕凿的用作栖息之地的摩崖洞穴，到代表建筑艺术雏形的“秦砖汉瓦”，再到象征着我国建筑艺术巅峰的明清建筑，中国建筑无不受到社会经济和文化发展的影响，呈现出朝代的痕迹和岁月的遗风。亳州的花戏楼（大关帝庙、山陕会馆）是一座馆庙合一、商神两用，具有经济、文化功能的明清建筑。因此，“它不是孤立的、脱离世俗生活、象征出世的宗教建筑，而是入世的、与世间生活环境联系在一起的宫殿宗庙建筑”[②]。而作为花戏楼精美建筑装饰代表之一的砖雕，其内容自然与时代和社会的变迁、经济和文化的兴衰以及民众的生活密切相关。

亳州花戏楼砖雕主要集中在“大关帝庙”（花戏楼）正面山门的墙面上。大关帝庙的正门（山门）是一座仿木结构的三层牌坊式建筑。建筑正面并列三个拱门，正门是行人出入的通道，左右钟楼、鼓楼各有一个拱门作为装饰，这两扇门只有在举行盛大仪式活动时开放。其砖雕图视以上坊立匾“参天地”和下坊横匾“大关帝庙”为中心向四周展开，布局端正有序、疏密得当、均匀对称。钟鼓楼门头上的砖雕布局同样方正规律，与正门的中心砖雕组块互相映衬，给人以协调美观的视觉效果。亳州花戏楼砖雕堪称艺术精品，是明清砖雕艺术的典范，以特殊的雕刻图像形式传递着文字语言难以表达的深层文化。它利用谐音、象征、联想、暗喻等手法向人们传达出“积极有为，奋发向上”的中华文化思想和“生生不息”的生命精神，体现了中华传统文化所蕴含的伦理道德精神和“天人合一”的

---

① 姜晓萍．中国传统建筑艺术[M]．重庆：西南师范大学出版社，1998．10．

② 李泽厚．美的历程[M]．香港：生活·读书·新知三联书店，2011．65．

审美诉求。

一、"生生不息"哲学艺术精神的观照

财富积累日趋雄厚、社会地位日趋上升的晋商群体,享受着明清时代商品经济带来的"国泰民安"的社会环境,崇尚儒家提倡的积极向上的人生观。"由于山陕会馆本身就是山陕商人为'敬关公、崇宗义、通商情、叙乡谊'而建立的场所,是山陕商人的精神家园,其雕刻内容具有丰富的文化内涵,处处蕴涵着当时深入商人内心的处世观念和经营理念,清晰地反映了清代儒商的意识形态。[①]"在花戏楼砖雕的选材和设计上,处处体现出晋商的这种精神诉求和文化审美心理。

花戏楼山门正中分布着三组砖雕,砖雕牌坊的最上坊是以一组《龙腾致雨》为正中,左右分列兜肚砖雕《鹰扬宴》和《鱼龙漫衍》的砖雕作品。将代表帝王形象的龙置于最高地位显示权威,表现出中华文化"九五之尊"的封建等级制。第二组围绕"参天地"匾额布局,是以《福禄寿三星高照》为上坊的砖雕序列,其长条兜肚《达摩渡江》、《老君炼丹》和对应挂芽《魁星点状元》、《文昌帝君》成为了整幅砖雕文化内涵的灵魂所在。它象征了中国传统文化在长期的发展过程中形成的"儒释道"三教合一的文化格局。第三组以匾额"大关帝庙"为主,包括匾额的上坊《吴越之战》和下坊《全家福·郭子仪上寿》。匾额左右兜肚为《三酸图》和《甘露寺·拜乔国老》,下坊《全家福·郭子仪上寿》左右兜肚为《魁星点状元》和《文昌帝君》。最下坊为长条形砖雕《松鹤延年》,最下坊门头饰为半圆形边饰《二龙戏珠》。

花戏楼砖雕序列突出了大关帝庙祭祀对象关羽"忠义神勇"的精神,叙述了儒释道文化观照下的人间百态和人生百味。序列中也有表现经世流传的中华文化传统美德"孝"的作品,充满道家艺术生命情趣的《四爱图》,借古喻今的历史人物故事、富含智慧和生活哲理的典故成语的砖雕作品,表达中华民族美好愿望的祥云瑞兽砖雕纹饰或小品等。整幅砖雕内容极其丰富,表达委婉含蓄,文化意蕴深远。

最上方一组以《龙腾致雨》为正中,左右分列兜肚《鹰扬宴》和《鱼龙漫衍》的组雕,是中华传统文化审美意象和哲学思想的高度概括和艺术表现。龙是中华

① 曹瑞林,曹峥. 晋商文化对明清会馆雕刻题材的影响——以社旗山陕会馆为例[J]. 大众文艺. 2011(22). 120.

民族的象征，是中国精神的物化体现。龙“为麟虫之首，能兴风致雨，以利万物。[①]”《龙腾致雨》砖雕作位于山门砖雕正中央最高的位置，雕刻着龙腾虎跃的场景和生机勃勃的大自然景物。《鹰扬宴》的涵义为，在清代科举的武科考试制度中，武科乡试成绩公布后翌日，考官和考中武举者共同举行的庆祝宴会，名为鹰扬宴。武科状元被比喻为“鹰”，寓意是“威武”。“鹰扬”寓意着考中武举者将会像威武的雄鹰一样展翅高飞，前途无量。此幅砖雕图宣扬一种积极向上的进取精神，鼓励人们积极生活，努力奋斗。同时，它也象征着晋商在其经济社会地位处于上升时期的乐观心态。《鱼龙漫衍》的砖雕取材自民间的百戏节目，表现了“鱼龙相嬉，跳跃漱水，雾障日毕，炫耀日光”的戏剧场景。龙是古代帝王的象征，它也代表着喜庆欢乐。“鱼”的谐音为“余”，如“年年有余”象征着年景好，丰稔昌盛。其寓意可理解为人的精神生龙活虎，生活场景生气勃勃，社会发展态势蒸蒸日上。整组砖雕呈现出一派风调雨顺、国泰民安的太平盛世，传达出积极向上的思想。它体现出中华民族积极乐观的生活态度和生生不息的生命精神。

中华文化能独立于世界优秀文化之林，持续发展数千年而延续不断，其发展的内在源泉在于中国儒家哲学提倡的“积极有为，奋发向上”的生活态度和“生生不息”的生命精神。孔子曰，“其为人也，发愤忘食，乐以充忧”。作为人，应该发愤努力，积极向上，乐观地对待现实生活。春秋时期，孔子就提出了一个流传至今的“不知生，焉知死”的哲学命题。基督教以“天国的幸福”和“上帝的召唤”来教化和抚慰“受苦难”的子民，宣扬不追求现世的幸福，忍受现世的精神和身体的痛苦，而幻想来世的快乐，这样死后就能进入一个并不存在的“天堂”。佛教虽强调现世更多做善举，但采取“酒肉穿肠过”的“不作为”态度，奉行“苦海无边，回头是岸”的哲学思想。“中国古代哲学有一个优良传统，即不重视死后的问题，不追求来世幸福，不将道德建立在灵魂不灭的信仰之上。[②]”《论语》记载：“子路问事鬼神，子曰：未能事人，焉能事鬼！曰：敢问死！子曰：未知生，焉知死！”《论语·先问》中孔子认为重要的是知生，而不是知死。人活在当今，其道德价值、生命追求都应建立在现实生活之上。要积极努力地奋斗，乐观地过好当下，“不知老之将至云尔”，不要考虑死后的事情，不要消极悲观地把希望寄托在来世上。

---

① 魏彪. 花戏楼的砖雕艺术[M]. 香港：香港天马图书有限公司，2000. 3.

② 张岱年. 文化与哲学[M]. 北京：中国人民大学出版社，2009. 218.

从战国时代起,中华民族就以"天行健,君子以自强不息"的思想作为激励民族奋发向上的精神动力。"健"是生命的本性。《周易·系辞传》说:"天地之大德曰生","生生之谓易",意思是天地的根本性质是"生生不息",流变不止。"天体运行,健动不止,生生不已,人的活动应当效法天,故应刚健有为,自强不息。[①]"这是一种积极健康、催人向上的生活态度,也是亳州花戏楼砖雕所传达的,蕴含在中华文化中的人生哲学和生活态度。这种精神在砖雕《龙腾致雨》、《鹰扬宴》和《鱼龙漫衍》里得到了形象的阐释和精准的再现。

二、"参天地"中"和"的文化审美意象

按照冯友兰先生的"哲学特性说",中国传统文化"具有以儒学为主导因素的哲学特性"[②],是以儒家思想为主,佛家、道家学说为辅的伦理文化。这种文化一直延续发展至今。花戏楼山门上的砖雕作品的内容选择和位置排列,相当于一种社会政治文化符号,很好地诠释了中国"儒释道"文化的历史格局和发展方向,体现了中国艺术的最高境界:"和"。最具代表性的是山门正中匾额上刻的"参天地"三个字,以及以《参天地》为主的砖雕序列,包括匾额上坊《福禄寿三星高照》,匾额两侧兜肚《夔一足》、《达摩渡江》和《老君炼丹》。此组砖雕序列图视虽少,但其凭借高度洗练的文化审美内涵,可被视为所有砖雕作品的核心和精华。这组砖雕序列浓缩了整幅砖雕图蕴涵的蔚为大观的文化内容,集中表现了建造者的文化价值观和审美观,寓意深邃,内涵丰富,把中国文化所追求的审美最高境界"和"体现的几近完美,具有强大的艺术震撼力。

《参天地》砖雕系列的排序,艺术地体现了中华文化"儒释道"三家并存共生的"和"的审美内涵。"参天地"三个字代表在中华文化中占统治地位的儒家文化,左右两侧分列"达摩渡江"和"老君炼丹"的砖雕故事,分别代表释(释迦牟尼)与道,即佛教和道教。这一序列砖雕位于大关帝庙匾额上坊,说明生活在封建社会的人民需要精神信仰。在人生理想和价值目标上,花戏楼砖雕表达的是儒释道三教合一构成的中华文化主体:儒家追求"内圣"、"尚德"的使命精神,道家追求"天地与我为一"的闲情逸趣,佛教追求如梦如幻的"超度"境界。三教以儒家文化为主,各有所长,各有发展,长期以来孕育着中国文化,滋润着中国人的

---

① 张岱年,方克立. 中国文化概论[M]. 北京:北京师范大学出版社,2013. 296.

② 程裕祯. 中国文化要略[M]. 北京:外语教学与研究出版社,2011. 10,97.

心灵,化解人生之忧。它们使人在孤寂无助中找到精神的寄托,在困惑痛苦中找到化解的途径,在混沌迷茫中求得心灵的平静,从而实现人与自我之间的“和”。在中华文化数千年的发展过程中,“儒释道”三种文化崇其所善,相辅相成,为中华民族提供了源源不断的人生智慧和思想源泉,成为中华民族赖以发展生存的文化基础和精神家园。“中国艺术自觉地追求表现天地之心,拟太虚之体,因而把‘和’作为最高境界。对‘和’的追求是艺术家通过自身对‘和’的基本精神的体会,用艺术的形式表现出来的。[①]”

先秦时期,学术繁盛,百家争鸣;汉代提出“罢黜百家,独尊儒术”。从此以后,儒学一直居于中国文化的统治地位。儒家文化是一种伦理文化,“德”为最高道德规范。而要达到“德”的规范,“和”至关重要。儒家推崇人与天地和、人与人和、人与社会和、人与自我和;要实现个体内部的和,个体与整体的和,整体与整体的和,如此种种,才能达到“和”的最高境界。从社会价值观的角度看,儒家尚道德,重人伦,讲仁爱,主张“和为贵”的中庸之道,这对于维护社会秩序起到了积极作用。道家崇尚自然的不法精神,通过否定一切外在形式的束缚,用人生智慧和艺术情趣来化解人生之忧,提高人的生命价值。佛家提倡自识本心,见性成佛,修炼自我,普度众生。从精神生活的层面看,儒家追求精神的自我完善,向往成为道德圣人。道教提倡精神的解放与超脱,向往“独与天地精神往来”的境界。佛教追求精神的净化超升,向往“涅槃”的境界。其实,道家与儒家殊途同归,最终都强调个人与无限的宇宙契合无间——“天地与我并生,万物与我为一”(《庄子·齐物论》)。道教和佛教都主张拓宽心灵,追求尽善尽美的精神境界,成圣、成真、成善,达到与自然、与自我、与他人“和”的境界。无论是儒家的秩序,道家的情趣,还是佛教的禅心,都对中国这样一个倡导“修身齐家治国平天下”的大一统国家有着重要作用和影响。

《参天地》组雕就是用象征手法,把《参天地》、《达摩渡江》、《老君炼丹》和《夔一足》组成一个序列,表明儒教、佛教、道教三教文化共处一片蓝天下,各引一端而又互补相生,达到了“和”的境界。福禄寿三星在中国文化中象征高官厚禄、福寿满堂。《参天地》两侧分列的是对称兜肚《夔一足》。夔是古代神话传说中的怪兽,“出入水则必风雨,其光如日月,其声如雷,其名曰夔”(《山海经·大

---

① 张岱年,方克立. 中国文化概论[M]. 北京:北京师范大学出版社,2013. 192,246,248.

荒东经》)。又传“夔”是舜的乐正(古代司音乐的官员)。古人认为音乐“乃天地之精也”,音乐能“平天下”,“故唯圣人为能,和”。因夔能和乐(懂音乐),即为圣人,“一夔足也”,有他一人足够。“夔一足”反映出尽人之性、人尽其才的思想,影射关羽忠义神勇,可与天地并列,有他一人足也。音乐为天地之精华,用音乐化育生命,化育精神,化育灵魂,与参天地的思想非常契合。“尽已之性,尽人之性,尽物之性,以参赞天地之化育。儒家博施济众、成己成物的仁心,道家万物与我为一的宽容,佛家普度众生的情志,都是天地人统一观念的结晶。①”

“参天地”确立了人在宇宙中与天地并列为三的崇高地位,体现了中华文化“天人合一”的宇宙观。“参天地”的思想来源于原始儒学。远古时期,古人就对天地有一种神圣的崇拜。中国文明属于农业文明,农民从事传统的土地耕种,以自给自足的小农经济为本。他们信仰宗教是为了祈求良好的自然环境,因为环境与农业生产有直接联系。于是,天地、日月、风火、雷电等成为先人崇拜的对象。在古人拜天地、敬鬼神、祭祖先的“三拜”中,对天地的崇拜是最为突出的。不仅百姓敬天拜地,连皇帝都要参拜天地。现在北京城留存的“天坛”、“地坛”即是古代皇帝参拜天地、祈求风调雨顺的神坛圣地。战国儒学思想家荀子的《礼论》和西汉史学家司马迁的《史记·书·礼书》里对这一思想都有阐述:“天地者,生之本也”,“无天地,恶生?”天地指的是人赖以生存的自然环境,是万物生存的根本,没有天地,人怎么生存?“故礼,上事天,下事地,尊先祖而隆君师,是礼之三本也。②”礼就是上拜天,下拜地,祭祀祖先,而后尊敬君师。古人认为,要想得到风调雨顺、五谷丰登的好年景,想要避祸求福,安享太平,就要对天地神灵顶礼膜拜。出于对大自然的敬畏和崇拜,要善待自然、遵从自然规律,与天地和,与自然和,通过与天地的和谐相处,达到“天人合一”的理想境界。

孔子在《至诚可参天地》中阐述了人与天地的关系:“唯天下至诚,为能尽其性;能尽其性,则能尽人之性;能尽人之性,则能尽物之性;能尽物之性,则可以赞天地之化育;可以赞天地之化育,则可以与天地参矣!”孔子认为,天下极至真诚的人,才能充分发挥他的本性;能充分发挥他的本性,就能充分发挥众人的本性;能充分发挥众人的本性,就能充分发挥万物的本性;能充分发挥万物的本性,就可以帮助天地培育生命;能帮助天地培育生命,就可以与天地并列为三了。

---

① 王宁.中国文化概论[M].长沙:湖南师范大学出版社,2000.228.

② 程裕祯.中国文化要略[M].北京:外语教学与研究出版社,2011.10,97.

花戏楼山门砖雕的“参天地”三个字位于正中，体现出中华民族对天地的崇拜之情，彰显出人对与自然和睦相处的渴望，显示出创作者对中华文化宇宙观和生命精神的深刻领悟。“参天地”代表人与天地并列为三，也象征着人与天地和谐相处、鼎足而立，能得天地之精神，完成生命理想；“以平等精神体察宇宙间一切存在的价值，完成生命”[①]。它象征着人们能认识到自我以外的天地，以顶天立地的自我与天地并列，这样，人的精神和地位就都达到了最高境界。“参天地”立于大关帝庙的山门上方，指的是“参天地”的人——关羽。人们认为他象征着忠义与武勇，能够永佑世人，可与天地相媲美。“参天地”的哲学思想无疑为晋商在商场上奋斗、在生活中打拼和实现自己的生命价值提供了精神支柱和内在能量。

实际上，中华文化中通常意义上的“儒释道”文化中，“儒”指的是以孔孟学说为主的儒家文化或儒教；“释”指佛教文化；“道”可以是道家或道教，或指二者的结合：道家指老庄，即老子和庄子创立的学说；道教则源自于东汉时出现的奉黄帝和老子为教主的黄老道，它流行于中国民间，吸收了道家宣扬的养生理论和长生不老的思想。“老君炼丹”的故事源自于道教而非道家，讲的就是太上老君炼制神丹以求成仙的故事。太上老君是道教最高神明之一，被奉为道教的鼻祖，是被神化了的历史人物。它的原型是先秦最著名的思想家之一，老庄学派的开创人老子。从这层意义上来说，道教和道家出自于同一人——老子。但是道教和道家的思想内容则是两个完全不同的概念。

道教与道家的基本信仰都是“道”，但道教注重从宗教的角度理解、阐释老子的“道”，认为“道”是宇宙万物之本原，同时又是“灵而有性”的“神异之物”。道教宣扬的是“得于道果，修炼成仙”。“仙”不但指灵魂常在，而且指肉体永生。修行办法有道术和道功。“道功指修心养性的内养功夫，道术是修命固本的具体方法。”黄帝和老子都认为，“只有清净无为，恬淡寡欲，才能体会‘道’”。而道家的代表人物老子和庄子所说的“道”，“是一个终极实在的概念，在本质上既不可界定也不可言说。不能以任何对象来限定，也不能将其特性有限地表达出来。它是不受局限的、无终止的一切事物的源泉与原始浑然的总体。[②]”道家崇尚的是“独与天地精神往来”的思想境界，主张“无为而不为”的处世哲学，提倡有艺

---

① 张岱年，方克立. 中国文化概论[M]. 北京：北京师范大学出版社，2013. 192，246，248.

② 张岱年，方克立. 中国文化概论[M]. 北京：北京师范大学出版社，2013. 192，246，248.

术情趣的人生态度,追求精神的自由超脱与解放。道教和道家,两者一个是宗教,一个是思想学说。但艺术表现形式往往是抽象的、象征性的,尤其是建筑装饰、绘画等艺术形式。一旦在此类艺术作品中,儒、佛、道三教圣人共聚一堂,那么三教之间的界限就会变得模糊,这在民间的祠堂、寺庙建筑装饰中尤为明显。

三、大关帝庙承载的忠义文化

明清时期,山陕商人建的所有会馆都是以关帝庙为主、馆庙合一的建筑。馆内供奉关帝的大殿称为大拜殿、关帝庙大殿或关帝殿等。“晋商行会多有自己行业所祭拜的偶像,如牲畜行供马王,纸行供蔡伦,以其作为联结同业者的纽带和共同的精神支柱;此外,没有一个商会不祭拜关帝,关帝是晋商共同崇拜的偶像。①”刻于亳州山陕会馆山门正中匾额上的“大关帝庙”四个字,说明建造者的初衷是建一个供商人们祭祀崇拜关公的关帝庙,而戏楼是后来增建的。“匾额是古建筑的眼睛”②,它将书法和建筑艺术融合在一起,不仅说明了建筑的名称和性质,还具有装饰作用,同时表达了建造者的情感。一个“大”字既表现出了建筑的恢弘大气,又含有对“忠义神勇关圣大帝”的崇拜敬仰之意。大关帝庙大殿里除关帝大圣的雕像外,有的还配有关帝的赤兔马雕像。馆内的匾额、楹联、墙雕中处处可见赞颂关帝忠义、诚信之词,雕刻、彩绘象征忠义的图案比比皆是。为何一个武将会受到商人的如此崇拜,以至于“无关帝不成庙,无关公不成馆”呢?关公精神与商业文化有何联系?

中华文明是信仰多神论的文明,渗透中国人生活的实质性宗教崇拜主要有三大崇拜对象:天地、祖先和君师。对天地的崇拜源于对生存环境的依赖,对祖先的祭祀出自人类血脉的延续,而对君师的崇拜,则来自内心的敬仰和精神上的需要。关公的“忠义仁勇”是中华传统文化的象征符号之一。中华文化中可为万世师表的圣人,传统意义上只有孔子和关公两个人。这一文一武两位圣人体现了中国治国的两大机制,是封建国家赖以长存的精神支柱。孔子是中国封建社会的著名教育家,儒学思想文化的代表,最高封号为“大成至圣文宣先师”,是一位圣哲。关羽是具有忠义勇武精神的武将和民间崇拜的保护神,最高封号为“忠义神勇关圣大帝”,被尊为道教神祇。人们在各地修建庙宇祠堂祭拜“二

① 孔祥毅.晋商商帮溯源.晋商研究[M].北京:经济管理出版社,2008.20.
② 朱广宇.图解传统民居建筑及装饰[M].北京:机械工业出版社,2011.132.

圣"。孔子为文圣,为此敬奉孔子的庙宇叫做文庙;关公为武圣,供奉关公之庙宇被称作武庙。实际上,关公在民间的影响力远远超过孔子,对关公的神祇崇拜成了民众生活的一部分。于是大大小小的关帝庙、关公祠遍及中国各地城乡,出现了"今且南岭极表、北极寒垣,凡儿童妇女,无有不震其威灵者。香火之盛,将与天地同不朽"的盛况。①

关公,名关羽(162～220 年),字云长,生于东汉桓帝延熹年间,山西解州人(今山西运城)。关羽本是三国时期蜀汉帝王刘备手下的一位武艺超群、性情刚烈孤傲的著名武将,死后被追谥为壮缪侯,即关壮缪。陈寿《三国志》称其"雄壮威猛",为"万人之敌",其性格"刚而自矜"。在三国时期的征战中,关羽对其主刘备的忠义赢得了魏国曹操的惜才之心,曹操意欲劝降之。但关羽恪守忠义,誓死不从,战斗中连斩曹操数员大将,但曹操依然对他宽厚以待。为报答曹操这份情义,在赤壁之战后曹操败走华容道时,关羽违反军令放走了曹操。关羽的这种忠义仁勇、诚信秉忠的精神受到民间的推崇,也得到历代君王的褒奖。他的品质与儒释道三大宗教联系密切,具有浓厚的宗教色彩。儒教尊奉关公为"忠义勇敢武圣人",希望天下人都能像关公一样以忠义待人,以公心待人,这样天下人就再也不会相互欺诈,四海之内皆兄弟。佛教把关公视为珈蓝菩萨,认为他能播撒男性的慈爱、忠信和仁义于普天之下。道教封关公为"五路财神"中的武财神和保护神,认为他能给人们带来财运。

关公崇拜在中国传统文化中具有重要地位和重大影响。关公崇拜始于宋朝,鼎盛于清代。明清两代关羽已经赫然成为国家神——"三教圆融"的关圣。关公崇拜遍及全国,关公也几乎成为了一尊宗教神祇。关公崇拜能够"与时偕行"、历六代而愈盛,源于历代皇帝的加封和民间活动的推崇。宋代崇宁元年(120 年),信奉道教的宋徽宗追封关羽为"忠惠公",六年后加封为"武安王",元天历八年(1328 年)加封为"显灵义勇武安英济王"。明万历三十三年(1605年),崇信道教的万历皇帝将关羽正式列入道教神祇,封为"三界伏魔大帝、神威远镇天尊关圣帝君"。清顺治九年(1652 年),清朝政府出于笼络汉族臣民的需要,封关羽为"忠义神武关胜大帝"。清代关庙中的一副对联颇能概括关公在中国封建社会的历史文化地位和巨大影响:"儒称圣,释称佛,道称天尊,三教尽皈

① 胡小伟. 关公崇拜溯源[M]. 太原:北岳文艺出版社,2009. 2.

依。式詹庙貌长新,无人不肃然起敬;汉封侯,宋封王,明封大帝,历朝加尊号。矧是身神功卓著,真所谓荡乎难名。①"经过一千多年,在上至皇帝、下至民间、兼及三教的共同推崇、不断塑造及《三国演义》小说的影响下,加之"神话传说和定期的仪式活动激励着百姓对关公保持虔诚的信仰",关公信仰得以不断延续,历经千年始终保持着其在民间的影响力。② 中国人心目中的关公,已从一位历史人物升华成为中华民族的一尊兼具道德和武勇的神祇偶像。

关羽祖籍山西。明清时,山西民众对关羽的信仰崇拜已经达到至高无上的地位,这种民俗文化已具有了宗教般的凝聚力。所以当山西人走出故土,在外谋生祈求神灵护佑时,关羽在他们心中便成为了"保护神"和道德楷模。中国的农耕经济造就了华夏民族"早出暮入、耕种树艺"的生活习惯,"重土难迁"的思想根深蒂固。中国历朝不渝奉行"重农抑商"的政策,认为手工业者为奴,商贾为贱民。而到了明清时期,商品经济开始发展,海内外贸易繁荣,使社会"仕农工商"的等级结构逐渐向"商仕农工"转变,金钱成为社会主宰和价值衡量标准。各地出现"以经商为第一等生业,科第反在其次者"③的风尚。而最初的山西商人走出故土、颠沛流离的一个重要原因是,山西地处黄土高原,土地贫瘠,自然灾害频繁。在清朝的两百多年里,自然灾害多至一百多次,最长的一次旱灾达十一年之久。人们流离失所,性命难保。由于山西"一方水土养不活一方人",山西人只好背井离乡"搭着命走西口","这一走就是几百年,走成惊世大商帮"④。《走西口》是山西家喻户晓的民歌,凄凉无奈而充满期望的歌声,表达的正是明清明期晋商艰辛创业的心路历程。早期的晋商创业者们远走他乡、浪迹荒漠、乘槎浮河,创业环境的艰险让人感到无助。面对命运的不可知和无奈,他们只能祈求神明庇佑。于是忠义勇武、威严仗义的关公被认为是招财进宝的财神和"保境立命"的保护神。"关羽的义脱离了庸俗和市侩的气息,他的品质能够激励人们在人际交往中讲义气,抵制荣华富贵的引诱。这符合封建经济中小生产者的心理,是下层人民所向往的脱离了利益的人际关系。他的勇武,更被理解为能救

① 胡小伟.关公崇拜溯源[M].太原:北岳文艺出版社,2009.3.

② 杨庆堃.中国社会中的宗教[M]. *Religion in Chinese Society—A Study of Contemporary Social Function of Religion and Some of Their Historical Factions*.上海:上海人民出版社,2007.157.

③ 冯天瑜.明清文化史[M].上海:上海人民出版社,2006.50.

④ 河南社旗县委宣传部.赊店文化[OL].2011.35.

助弱小、惩治邪恶的神力。①”

“明清晋商称雄商界500年,是依靠一种精神信仰与理念的支撑,这种精神信仰就是被神化了的关公,这种理念就是独特的关公文化。②”山陕两大商帮的联合使西商强大起来,他们的经商足迹遍布全国各大商贸集镇和港口码头。他们在各地捐资修建山陕会馆,建关帝大殿祭拜关羽,对推进忠义文化影响很大。中华文化由此形成了“三教尊一人”的特殊现象。晋商修建大关帝庙祭拜关羽,是因为看重关羽的忠义和诚信。从此忠义诚信被当作经商做人的道德标准,被用来审视自己、约束他人、规范行业。与此同时,儒家“学而优则仕”的观念悄然变化,一些儒家贤士进入经商行列。儒家文化的核心是“仁”和“德”,孔子曰:“君子义以为上。”孟子在讨论生与义的问题时,认为“生是重要的,义也是重要的;如果两者不能两全,应舍生而取义”。他说:“生亦我所欲,义亦我所欲,两者不可得兼,舍生而取义也。③”孟子这种“舍生取义”的思想,也正是晋商精神与儒家思想两种文化的契合点,对于中华民族精神的形成有着特别重要的意义。儒家文化的精髓渗透在晋商的做人处事、经商管理和生活追求等各个方面,晋商文化中处处可看到儒家思想的影子。

“晋商不仅把宗教伦理中的入世精神化为强大的实践理性,而且将关公的品质提升为塑造商业伦理的精神准则。④”关公崇拜文化内涵丰富,关羽精神中的“义”是中国的传统美德,也是晋商最为敬重崇拜关羽之处。修建大关帝庙的意义就在于让后人牢记商场上要“义”字当先,以义去“举其言、践其行”。大关帝庙中的“义冠古今”、“仗义秉忠”楹联扁额等,都突出一个“义”字。商人经商讲究和气生财,正所谓“买卖不成仁义在”。“义”含有“德”之意,晋商提倡做生意要诚实守信,“义中取利”,不发无义之财,不见利忘义、不欺诈、不缺斤短两等,也就是讲究公平竞争,是“和为贵”思想的体现。义指义气,如兄弟之义、朋友之义等。“义”成为了山陕商人在经商和做人方面的道德准则,他们的仗义和大方是一般人很难做到的。在山陕商帮中流行这样一种做法,如果一家店欠另

---

① 侯利敏.山陕会馆中的神祇崇拜——以社旗山陕会馆的关公崇拜为例[J].黑龙江史志.2008(14).45.

② 张正明.明清晋商与关公文化晋商研究[M].北京:经济管理出版社,2008.26.

③ 张岱年.文化与哲学[M].北京:中国人民大学出版社,2009.188.

④ 董立清,卫东海.关公信仰与明清晋商精神的发展[J].中共山西省委党校学报.2010-08(4).112.

一家店一大笔钱,到期无力偿还,借出店往往为了照顾欠账人的自尊心,就让欠账人象征性地还一把斧头、一个箩筐,哈哈一笑,就算了事。它反映了山陕商人的大方仗义,同时也说明即使在困境中,人也要讲究信义,不能有背信弃义的行为。比起莎士比亚笔下威尼斯商人夏洛克的贪婪自私,中国晋商的这种信义显得难能可贵。晋商的种种义举和传说让人们改变了对商人的偏见,扭转了世人眼中"无商不奸"的印象。

晋商崇敬关公"忠义"精神的另一个原因是商业上的需要。中国是一个宗法专制的社会,自古就有"以德治国"的思想。商家都强烈地意识到,信义是立业之本。在生意场上晋商奉行"诚信为木"的经商理念。这就是要求商人取信于民、取信于社会。关羽所代表的忠义、诚信成为晋商规范行业行为、守护商业繁荣的最高道德标准。因而晋商在商业活动中都能自觉地守望良知、群体致诚,创造了驰骋中国商海500年的历史。而现在,"诚信为本","义中取利"仍然是市场经济体制下,我国商人经商的道德标杆和取胜之本。关公的忠义与儒家提倡的诚信有惊人的相似之处。"在人伦关系中,中华民族不仅有诚信的美德,还有'报恩'的德性。'报恩'即知恩图报。报恩既是中国人的传统美德,也是道德生活的重要原理和机制。[①]""滴水之恩,当涌泉相报"在社会生活中是公认的美德,是"义"的重要内容之一。

花戏楼砖雕用比喻、象征等多种形式宣扬"仗义秉忠"和"知恩图报"的思想,如"大关帝庙"匾额上方的《吴越之战》、右侧的《甘露寺·拜乔国姥》,左上方的《衔环之报》等故事和神话传说,都有警世训喻的教化之意,告诫世人要讲信义、知恩图报,报父母养育之恩、长辈栽培提携之恩、朋友相助之恩等。忘恩就是负义,是不道德的。"知恩图报是中国人道德良知的重要组成部分,是中国道德质朴性的表征。[②]"而位于"大关帝庙"匾额之上的《吴越之战》则形容商场如战场,胜败乃兵家常事,要学会韬光养晦,经得起失败、挫折,才能取得最后胜利。同时告诫商人们要像该故事中的范蠡那样,不慕虚名,不恋钱财,取之于社会,回报于社会。[③]

---

① 张岱年,方克立.中国文化概论[M].北京:北京师范大学出版社,2013.214,215.
② 张岱年,方克立.中国文化概论[M].北京:北京师范大学出版社,2013.214,215.
③ 亳州市文联,亳州市旅游局.亳州之旅[M].北京:中国文化出版社,2009.92.

四、刻在青砖上的中华传统文化

(一)经世流传的儒家"孝文化"

亳州花戏楼山门砖雕组图序列中,匾额"大关帝庙"的下坊是砖雕经典作品《全家福·郭子仪拜寿》,因其祝寿场面之大、刻画人物之多,成为整座砖雕牌坊的代表作和精华。《全家福·郭子仪拜寿》整幅砖雕作品刻有42个人物。人物形态各异,满面春风,喜气洋洋。老寿星郭子仪端坐正堂,胡须垂胸,慈眉善目,面带微笑,和蔼可亲,身后的"寿"字清晰可见。他周围是众多前来祝寿的朝中文武官员。砖雕图的两侧亭台楼阁玉立,车马人流熙攘,一派富足祥和的景象。这一幅欢乐祥和、尽显天伦之乐的祝寿场景,表现出中华文化十大传统美德之首的"孝悌"之德,传达的是中华文化中孝文化信息,体现出中华民族祈求家庭和美、父慈子孝、延年益寿的美好愿望和儒家文化所追求的"和"的最高境界。

因平定叛乱而成为功臣的唐太师郭子仪,成为了中国民间最受推崇的"福禄寿全"的代表,集多子多福、高官厚禄和长寿于一身,成为中国文化艺术作品的原型。据说安徽省博物馆内收藏的一幅《郭子仪拜寿》砖雕图是该题材的作品中最为完整而最具代表性的艺术珍品,而《郭子仪拜寿》的场景也以砖雕、木雕、石雕和彩绘等不同的艺术形式出现在中国许多大型建筑装饰中,显示出中国人对孝道的推崇。

中华文化历来就有"百善孝为先"的道德认知。孝是建立在仁爱基础上的。中国历代思想家都认同仁与爱。孔子提出"君子义以为上"(《论语·阳货》)。"好仁者无以尚之"是《论语·里仁》的命题,认为道德是至上的。"义指道德原则,义的内容就是仁,仁是最高的道德规范。[①]""仁发端于人类集体生活中所形成的'恻隐之心',即同情心,它基于家族生活中的亲情。[②]""仁德"的核心是爱他人,爱父母长辈、爱兄弟姐妹、爱亲朋好友、爱天下人。仁的根本是孝悌,"孝悌也者,其为仁之本欤"(《论语·学而》),就是说要尽孝道,报答父母的养育之恩,赡养父母,尊敬老人。中华美德以向善的信念为主,主张自主自律,注重诚信的品德,首先就是要回报父母,对其他人也要知恩图报。成语故事中"乌鸦反

---

① 张岱年.文化与哲学[M].北京:中国人民大学出版社,2009.184.

② 张岱年,方克立.中国文化概论[M].北京:北京师范大学出版社,2013.213.

哺”和“羊羔跪乳”告诫人们要感谢父母的养育之恩,要有赡养长辈的孝心。“乌鸦反哺”的故事出自李时珍的《本草纲目·禽部》:“慈乌:此鸟初生,母哺六十日,长则反哺六十日。”意为小乌鸦在母亲的哺育下长大,当母亲年老体衰,不能觅食时,小乌鸦反过来自己寻找食物并喂到母亲的口中,回报母亲的养育之恩,直到老乌鸦死去。“乌鸦反哺”是乌鸦这种群居动物的生活习性。近几年,国外动物学家在观察研究群体生活的乌鸦时,确实发现了这种“养老”的行为,而在其他群居的鸟类中却没有发现这种情况。由此可见,“反哺”很可能是乌鸦所特有的一种社会性行为。

中华传统伦理文化的三纲五常中,“五常”指“仁、义、礼、智、信”,它一直是规范个人和社会行为的最高道德标准。清代官员邓钟岳(1674~1748)写的一纸案文精辟地解释了中华文化中的“五常”:“鹁鸽呼雏,乌鸦反哺,仁也;鹿得草鸣其群,蜂见花聚其众,义也;羔羊跪乳,马不欺母,礼也;蝼蚁塞穴以避水,蜘蛛罗网以为食,智也;鸡非晓而不鸣,燕非春而不至,信也。禽兽尚有五常,而人为万物之灵,岂无一德乎!”《孝经》曰:“天地之间人为贵。”荀子说:“水火有气而无生,草木有生而无知,禽兽有知而无义,人有气,有生,有知,亦且有义,故最为天下贵。[①]”连动物都能做出如此感人的善举,人为何不能呢?中华民族是一个家庭观念非常强的民族,中国人认为孝敬父母是一种重要的社会责任。人们心理上的“反哺情结”是维系社会与家庭和谐和安宁的重要力量。在中国,不孝被认为是大逆不道,不孝者会因破坏社会和谐和秩序而受到社会道德的谴责。

花戏楼砖雕序列中宣扬“孝文化”的还有钟楼门头上枋,紧邻“钟楼”匾额左右两边的《蟠桃孝母》和《蒸黎(梨)休妻》雕饰。“蟠桃孝母”的故事来自于中国民间的神话传说:一白猿甚孝顺,一日,其母病愈想吃蟠桃,猿去孙膑处偷桃被捉。猿跪泣说出缘由,孙膑视其为孝子,赐桃后放回。这个故事说明,孝道受到社会的普遍赞同和践行。“蒸黎(梨)休妻”讲的是孔子四大弟子之一的曾子的故事。曾子以孝著称,其妻因给母亲蒸的梨不熟而遭到休弃。曾子的举动从现代社会的价值观看来似乎不可思议,但它真实地反映出“孝”在中华民族的道德观中所占的地位。

位于鼓楼门上枋,“鼓楼”匾额左右两侧的砖雕图是《燕山教子》和《王质烂

---

① 张岱年.文化与哲学[M].北京:中国人民大学出版社,2009.175.

柯》。“燕山教子”说的是窦家父亲高义笃行、家法严明、教子有方,故“父慈子孝,兄友弟恭”,最后五子登科的故事。“燕山教子”是中华文化中完美家庭教育的典范。自古以来中国就有严父出孝子的说法。它强调了为人父母者对构建和谐社会所应尽的家庭责任,体现了中国人根深蒂固的家庭观念。而“王质烂柯”则通过神话故事告诫世人,尤其是青少年:一寸光阴一寸金,寸金难买寸光阴。时光荏苒一去不复返,不要游手好闲、无所事事,要爱惜时间,珍惜生命。两则故事都充满了对后代的期盼,鼓励人们只要努力奋斗,就能走上仕途,有所作为。

古代中国是一个等级制度森严的社会,封建伦理道德的“三纲五常”规范着人们的日常行为,影响着人们的实践活动。三纲即“君为臣纲,父为子纲,夫为妻纲”,意思是为臣、为子、为妻的必须绝对服从于各自的君主、父亲和丈夫,同时君、父、夫也必须为臣、子、妻作出表率。它规范了封建社会中君臣、父子、夫妇之间的道德关系,由此产生人际关系的“五常”:仁、义、礼、智、信。这是儒家文化提倡的一种人伦“秩序”,其核心是“礼”。花戏楼砖雕不仅在故事内容上,在作品的布局安排上也体现出“礼”的规范思想。

中国人安排家人居住的位置时也要按照地位、身份的高低和年龄的长幼来分配,以体现出对长辈的孝心和礼数。譬如,坐北朝南的房屋以东为上,因此长辈或老人住在东屋;以西为下,所以晚辈或子孙们住在西屋。久而久之,这种分配方式形成了一种习俗,成为了“礼”的组成部分。花戏楼砖雕位置的安排就体现了这种中华传统文化中的“礼”。如东侧钟楼是体现“孝文化”的两组砖雕《蟠桃孝母》和《蒸黎(梨)休妻》,而与西侧鼓楼对应的是《燕山教子》和《王质烂柯》。钟楼在东侧,按礼的规范,东侧是长辈居住之处,因此两组体现“尽孝”的砖雕被安放在东侧;而西侧鼓楼对应的砖雕反映的则是“教子”的思想内容。

儒家文化的“中庸之道”、“立德正心”思想在建筑文化中的体现就是建筑“不正不威”以及“居中为尊”的理念。中华传统建筑文化还非常讲究建筑平面布局以中线为轴左右对称,这也是儒家“中正有序”思想在建筑上的具体体现。以中线为轴对称,符合人们的审美观,也是自然界中动植物自身构造形式的一般规律,“对称的人和物具有安定感,与人类祈求安稳、平衡的心态相吻合”[1]。在有一定规模佛教寺庙建筑中,钟楼和鼓楼都是相对称而建的。花戏楼最初是用

---

① 朱广宇. 图解传统民居建筑及装饰[M]. 北京:机械工业出版社,2011.4.

于祭拜关公的关帝庙,建有钟楼和鼓楼,钟楼在正门的东侧,鼓楼在正门的西侧。据说唐代以前寺院只建钟楼,不建鼓楼。我国早期寺院为了讲究对称,也有在今鼓楼位置建楼的,但只是作为保存佛典之用。后寺院内藏经渐多,另置藏经楼于大殿之后,原作藏经之用的亭阁式小楼则改为鼓楼。钟楼和鼓楼的位置安排还依照"晨钟暮鼓"的次序。"晨钟暮鼓"是我国寺院的一种传统。"钟、鼓声是僧团内集合和集体活动的信号,后来应用范围逐渐扩大,钟、鼓成为佛教法事甚至社会礼仪活动中不可缺少的器具。鼓楼之鼓与钟楼之钟配合有序,早晨先敲钟后敲鼓,晚上先敲鼓后敲钟,故名'晨钟暮鼓'。[①]"所以,中国古代所建造的规模较大的寺院庙宇,都在建筑物的左面建有钟楼,右面建有鼓楼。

(二)"崇儒尚文"的晋商文化

历史上,晋商和徽商都被称为儒商。"在儒家思想文化影响下从事商业活动的人称作儒商,而晋商就是儒商的典型代表。晋商以儒家文化和道德规范自己的行为方式、经营之道和价值取向"[②]。花戏楼砖雕中浸透了儒家文化的精髓,随处可见体现儒家"义"和"仁"思想的作品,反映出晋商"既崇商更尊儒尚文"的事实[③]。山陕商人建大关帝庙祭拜关羽,就是晋商尊儒最明显的表现。

"晋商文化涉及的内容很广。从商业伦理、经营理念、创业之路、管理之道、企业文化、产业思想等方面看,其特点可归纳为六点:一是关公崇拜;二是唐晋遗风;三是地缘贸易;四是乡土轴心;五是人本思想;六是官商相护。[④]"关公一生具有"忠义"、"诚信"的品质,是忠实地实践儒家提倡的仁、义、忠、信的典范。他的忠义勇武的品质体现在晋商的经商理念中,就是"诚信为本"、"义中取利",这两点被奉为治商之道,成为晋商获得成功的根本原因所在。在晋商看来,义是为人处事的根本,更是经商的根本。"利从义出,先予后取"是晋商经商理念的体现,说明儒家文化已深入晋商的思想。清代祁县富商乔致庸把经商之道总结为"守信、讲义、取利"。乔家的复字号兴旺百年,成为信誉的代名词,源于乔氏复字号复盛油坊的一次"胡麻油事件"。乔氏复字号复盛油坊的通顺店从包头运大批

---

① 中国佛教文化网[OL].2008-01-18.

② 贡振羽.从儒家思想看晋商文化[J].山西高等学校社会科学学报.2008(12).30,31.

③ 赊店历史文化研究会.赊店——中国历史文化名镇[M].郑州:大象出版社,2005.212.

④ 孔祥毅.晋商商会与晋商文化的发展.晋商研究[M].北京:经济管理出版社,2008.82.

胡麻油往山西销售，经手店员为贪图厚利，在油中掺假。店主乔致庸得知此事后，宁可失一时之利，也要挽回商誉。他命人连夜在全城贴遍告示，说明通顺店掺假的事件及其原因。同时，他承诺凡是近期到店买过胡麻油的顾客都可以全额退货，以示赔罪之意。商号虽蒙受了损失，但因店主诚实无欺，义中取利，复字号的油更受青睐。在亳州的晋商也秉承了这种理念，并利用各种途径，包括借助砖雕艺术来大力宣扬这种诚信为本，义中取利的儒商文化精髓，告诫他人要遵循"义利兼济"、"义为上"的理念，并以此自勉。

晋商重商是毫无疑问的，但其"尚文"的理念在花戏楼砖雕图视上的表现也很突出。在花戏楼山门上的52幅砖雕中，带有明显"崇儒尚文"思想的作品就有20幅。如位于戏楼正门门头上方砖雕组块中的挂芽《魁星点状元》和《文昌帝君》，兜肚《天禄书镇》和《麟吐玉书》就是非常典型的例证。魁星在中国文化中被称为主宰文章兴衰的神。魁星的形象往往是右手握一枝大毛笔，又称朱笔，用来点定中考人的姓名。因古代的科举考试都是以文章的优劣和文笔的好坏来衡定考试者中取与否，所以，魁星的权利及形象都令文人敬畏和崇拜。而文昌帝是民间和道教尊奉的掌管世人功名利禄之神，又称文曲星，也为读书人崇拜。雕饰中出现的魁星和文昌帝君，往往分别脚踩大鳌鱼和白虎头。魁星朱笔一挥点状元，文昌帝君文质彬彬，手持中举者的名录。这幅砖雕的寓意是让读书人发奋读书，将来一定会独占鳌头，大展宏图。现在市场上出售以"文曲星"为商标的电子辞典，说明了现代读书人仍然把"从文"与对"文昌帝君"的崇拜联系在一起。花戏楼砖雕工艺匠将这两位司文的神明的像放在突出的位置，说明晋商对文帝的崇拜和敬畏之情以及对于功名利禄的追求向往。同时，这四幅作品被置于门头左右上方较为明显的位置，思想内容相似，布局对称，四角构成一个系列组块，有很强的艺术效果。它改变了人们对商人"一身铜臭凑不识文"的偏见，说明晋商既重商，又崇儒尚文。

中国封建社会几千年来一直都有"万般皆下品，唯有读书高"、"书中自有黄金屋"的崇文思想。中国人教育子女从小就要发奋读书，"两耳不闻窗外事，一心只读圣贤书"，要"头悬梁、锥刺股"，以便"读得十年寒窗苦，求得一日功名就"，日后加官晋爵，前程远大。这种"学而优则仕"的思想源于中国封建社会的人才选拔制度"科举制"。科举制指设立考试科目，以考试选举有见识、有才能的人。科举考试分为两种，每年都举行的考试科目叫"常科"，皇帝临时设立的科目叫"制科"。常科考生大体有两种：一是"生徒"，即在校学生，一是"乡贡"，即社会考生或历届生。中国历史上,在科举制度形成以前,依靠血统或裙带关系用人的“世卿世禄制”或“客卿制”是统治者的用人之道。那时,选士和任命权主要为豪门显贵所把持,选取标准以门第为重,于是常出现“上品无寒门,下品无士族”的现象。南北朝时期,豪门子弟凭借自己的显贵门第,就可以“平流进取,坐至公卿”。从隋朝开始,中国统治者选拔人才的制度开始变革。隋开皇七年(587 年),隋文帝废九品中正制,规定采用考试的方法选拔官吏,并于开皇八年设立“志行修谨”(有德)和“清平干济”(有才)两科以选拔人才。在此基础上,隋朝大业三年(607 年),隋炀帝设立“明经”、“进士”两个考试科目,以策问的形式选拔人才。据唐人杜佑《通典》十四卷载,炀帝“置明经、进士二科”,这是科举制度的雏形。唐代继承、发展和完善了始于隋朝的科举制,并使其逐渐发展成熟。隋唐以后,考中科举成为社会各阶层人民改变命运,提高个人及家庭的社会和经济地位,获得高官厚禄,追求荣华富贵的一大重要途径。

“‘商人’原指商朝人。商朝重工业、尚贸易,商朝人个个深谙贸易之道,人人精通渔利之法。[①]”在夏灭商以后,商人失去家国土地,只得东奔西走靠做生意维生。在周人眼里,商人成了“买卖人”、“生意人”的代名词,后衍生出“商业”、“商品”等相关词语。但在中国历史上,统治者对商业和商人一直持有偏见,“重农抑商”是明清以前的经济指导思想。明清时期商业贸易兴起,善于相时而动的山西俊杰多加入经商行列,“士农工商”从而转变成“商士农工”,因而商人就成了四民之首。贵族子弟从小读书识字,接受良好的儒学教育然后投靠商号,使得当时山西商界活跃的大都是儒学精英。在普遍崇尚“学而优则仕”的封建社会,山西人践行的则是“学而优则商”。

花戏楼(大关帝庙)的建造年代——清顺治十二年(1656 年)正是科举制度兴盛的时期。尽管山陕商人身在商界,但仍尊崇儒家文化。从晋商的价值取向和精神追求,以及当时的商业背景来看,晋商“尚文”也许是为了更好地从商。关于当时山西民众重商之俗,“山西巡抚刘与义向雍正皇帝上奏折说道:‘山右积习,重利之念,胜于重名。子弟之俊秀者,多入贸易一途。’清人刘大鹏说:‘当此之时,凡有子弟者,不令读书,往往俾学商贾,谓读书而多困穷,不若商贾之能致富也。是以应考之童不敷额数之县,晋省居多。’[②]”从现代理念来看,“崇儒尚

① 何川江.风雨商——中国商人五千年[M].北京:中国民主法制出版社,2009.24.

② 贡振羽.从儒家思想看晋商文化[J].山西高等学校社会科学学报.2008(12).30,31.

文”和“学而优则商”并非相互矛盾。“崇儒尚文”把儒家思想的诚信、仁义和忠恕精神引入商界，从而培养出具有儒学文化底蕴的商人群体，使儒学观念与经商理念互相融合，确保了晋商商业团体成员的精英化，从而使商业能够良性发展。晋商创立的选拔人才之道、经营理念和管理模式已被现代企业的实践证明是非常有价值和值得借鉴的。

五、《四爱图》折射的道家艺术生命精神

花戏楼砖雕《四爱图》指的是山门的中门两侧，在“大关帝庙”匾额左右对称而列的《王羲之爱鹅》、《周敦颐爱莲》和《鲁隐公观鱼》、《陶渊明爱菊》四幅砖雕。中国历史上，众多文人墨客不仅用文学和其他艺术形式为后人留下宝贵的精神财富，他们的许多生活轶事、兴趣爱好也折射出他们高尚的品格和高雅的艺术生活情趣，为后人世代传颂。四爱图中的四位文人贤士中，王羲之是东晋著名书法家，周敦颐是宋代哲学家，鲁隐公是西周鲁国的国君，陶渊明是东晋著名的文学家和隐士。这四幅砖雕即取材于中国历史上这四位名人的生活轶事。砖雕图案精美，雕工细腻，人物形象逼真，充满生活情趣，立意高远，耐人寻味。

《王羲之爱鹅》砖雕作品取材于我国晋代大书法家王羲之的生平故事。王羲之(321－379年或303－361年)，字逸少，号澹斋，原籍琅琊临沂(今属山东)，后迁居山阴(今浙江绍兴)，官至右军将军、会稽内史，是东晋伟大的书法家，也是中国最伟大的三大书法艺术家之一。他被后人尊为书圣，他的行书被誉为天下第一。王羲之一生酷爱书法，也喜爱观鹅养鹅。到了晚年，他离开喧闹的京城，来到风景宜人的江南水乡绍兴。在那里他经常漫步水乡泽园，观看水中浮游嬉戏的鹅群。

王羲之从羽毛纯白、体态优雅的鹅的动作和姿势中领悟到了书法艺术语言和运笔技巧，“观察鹅的划水、扑打、进食、嬉戏等动作并详加揣摩，将之融进书法之中。”[①]通过观察揣摩鹅群的动作姿态，王羲之将名为“鹅姿”的创意性运笔走势融入了自己的书法艺术，使其独具一格。在东晋永和九年三月三日，王羲之与四十多位文人在浙江会稽的兰亭聚会，“流觞曲水”，兴致勃发地在蚕茧纸上写下了“天下第一行书”《兰亭集序》，被认为是中国书法行书艺术的绝世之作。

① 杨耿. 苏州建筑三雕：木雕·砖雕·石雕[M]. 苏州：苏州大学出版社，2012. 118.

《兰亭序集》中有20个不同写法的“之”字,形态犹如鹅的姿态,“或埋首理翅,或引颈前趋,或昂首远顾,生动活泼,情趣盎然。①”清代著名书法家包世臣评价他的书法:“全身精力道笔端,定台先将两足安。悟入鹅群行水势,方知五指用力难。”(《晋书·王羲之传》)王羲之爱鹅养鹅,也留下了许多故事。“王羲之书成换白鹅”讲的是他为道士抄写《黄庭经》来换取鹅群的故事。后人又称《黄庭经》为《换鹅贴》。成语“入木三分”也源自于他的故事。据说有一次,他把字写在木板上,拿给刻字的人照着雕刻。雕匠用刀削木板,却发现他的笔迹印到木板里面有三分之深。这就是“入木三分”这个成语的由来。

《王羲之爱鹅》的砖雕作品中,艺术家观鹅的场景使人很容易联想到“白毛浮绿水,红掌拨清波”的诗句。艺术家专心致志观鹅,仿佛忘却了人世间的烦恼,进入了安宁自由、入神入境的“忘我”状态。王羲之爱鹅不完全是出自于兴趣爱好,同时也是他书法创作的需要。书法的魅力源自于中国汉字的优美结构和其象形表意的特征及其蕴藏的文化意境。书法作为一门艺术,反映了艺术家的个人风格,正所谓“字如其人”。“钟繇书如云鹄游天,群鸿戏海,行间茂密,实亦难过。王羲之书字势雄逸,如龙跳天门,虎卧风阙,故历代金之;蔡邕书骨气洞达,爽爽如有神力;韦诞书龙威虎振,剑拔弩张。”(萧衍《古今书人优劣评》)②王羲之最为著名的书法作品,流传千古的《兰亭集序》,其书法“中峰起转按提,以豪为之,线条如行云流水,字体结构极尽变化,风流潇洒之至。③”该作品展现出“气韵生动”之美,体现了出书法家率真、豪气、潇洒的性格,同时又展示出“超然物外”、“肇于自然”的气势,达到一种“天地与我同在”的道家精神境界,这也是书法艺术所追求的最高境界。

《周敦颐爱莲》砖雕图,形象地表现出宋代理学先师周敦颐欣赏莲花高洁正直品格,并用一生践行其品格的历史故事。周敦颐是北宋官员,一生清廉正直,不媚权贵,清高不凡。他的《爱莲说》中的“予独爱莲之出淤泥而不染,濯清涟而不妖”成为了千古名句。他以莲比喻君子,称“莲,花之君子者也”。君子应具备莲花的高洁品格,不入俗流,但又不能自恃清高。他自称独爱莲花高洁挺直,出水妙善,赞美莲花的与众不同、卓然独立的典雅和出淤泥而不染的高贵品格。无

---

① 赵静.中国赊店山陕会馆[M].郑州:中州古籍出版社,2013.111.

② 张岱年,方克立.中国文化概论[M].北京:北京师范大学出版社,2013.185.

③ 张岱年,方克立.中国文化概论[M].北京:北京师范大学出版社,2013.185,184.

独有偶的是,儒家文化也针对个人品格修养提出了“慎独”的方法。“所谓诚其意者,毋自欺也。如恶恶臭,如好好色,此之谓自谦。故君子必慎其独也。①”这是指独处时,要保持洁身自好,要自律而意念诚实。莲花原是佛教圣物,象征佛法和纯洁。常见佛或观音“踏莲”或“坐莲”,可见莲的地位之高。而在周敦颐笔下,通过对莲之意象的窃取与转化,成功地将原属于佛教的莲,变为儒家之莲、道学之莲。② 想要在这个繁杂喧嚣的社会“出淤泥而不染”,仅仅保持儒家提出的“慎独”远远不够,要有莲花那样身处污浊环境而能保持清廉高洁的君子精神。周敦颐描述的莲之高雅更接近于道家思想,即面对世俗功利不入俗流,而追求高雅、清静的生活。世界之大,无奇不有,坦然面对世界象征着人生的豁达和超然,是道家自由放飞心灵以求精神超脱的途径。

《陶渊明爱菊》砖雕是根据东晋大诗人陶渊明的隐居生活而作。陶渊明曾任江州祭酒、镇军参军。他志趣高远,不满现实,不慕名利,任职后不久便辞官而去,过起了隐居田园的生活。他一生钟爱菊花,被誉为“菊痴”。归隐后东园菊圃成了他耕锄作息之地,也成为历代文人吟诵的“桃花源”。他寓情于菊花中,其著名诗句“采菊东篱下,悠然见南山”反映了他回归自然后的生活状态,呈现出他淡泊宁静的心志。诗句“三径就荒,松菊犹存”(陶渊明《归去来兮辞》)描写了他自己的生活。小径荒芜了,但松树和菊花依然安静地存在着。诗人抛弃荣华富贵,避开车马喧嚣,在悠然自得的田园生活中获得了内心的自由和宁静。陶渊明之所以爱菊,主要是因为菊花象征了他的人格。正如周敦颐的评价:“予谓菊,花之隐逸者也。”萧瑟的秋天,菊花傲霜怒放,正如陶渊明傲世而立。与其他隐士不同的是身为士大夫,他不仅能“超越私欲,摒弃奔竞媚俗”,还过着“夫耕于前,妻锄于后”的农夫生活而自得其乐。这是他最鲜明的隐士特征,“独与天地精神往来,而不傲倪于万物(《庄子·天下》)。正是在与自然的契合中,他的人格达到了一种逍遥之境”,而逍遥境界历来是道家内在境界的表现。

《四爱图》中描绘的前三位都是中国历史上著名的文人墨客,只有最后一位鲁隐公是古代国君。“四爱图”的创作者为何会选择“鲁隐公观鱼”作为四爱之一呢?纵观鲁国在历史上的地位可略知一二。鲁国第一任国君姬伯禽是周公姬旦的长子。周公曾为周朝定立所有的规章礼仪,后退隐于鲁国。鲁国被特许了

---

① 孙雨嘉.论周敦颐对儒家“慎独”思想的超越[J].湖南行政学院学报.2011(2).89.

② 郭初阳.《爱莲说》:出水妙善[J].教师博览.2011(1).6.

可以“世祀周公以天子之礼乐”的特权,故而成为与周朝中央政府最亲密且最有地位的诸侯国。在诸侯国议事等活动中,各国的排列次序也都遵循“尊尊而亲亲”的宗法精神,“周之宗盟,异姓为后”(《左传·隐公十一年》),即在姬姓诸侯国中,鲁国位于首位。用《国语·鲁语上》的说法,就是“鲁之班长”。鲁国是所有诸侯国中周礼保留得最完整的“礼仪之邦”,鲁国的国史也是最完整的。中国儒学鼻祖孔子出生在鲁国,流传千古的《春秋》就是孔子根据鲁国国史改编而成的。由于《春秋》记载的内容开始于鲁隐公元年(前722年),鲁隐公也在史学上出了名。《周礼》与孔子都对中国几千年的文化产生了重要的影响,鲁隐公的故事很具有代表性。

鲁隐公(前722~712年在位),姓姬,名息姑,是西周鲁国第十三代国君。他为什么叫隐公呢?在中国古代,国君被称为“公”,有尊敬之意。而“隐”是谥号。古代皇帝的称呼往往和年号、谥号和庙号联系在一起。谥号是给死去的帝王、大臣、贵族或其他地位很高的人的称号,带有评判性。谥号始于西周,废止于秦朝,在西汉再次兴起。周公旦和姜子牙有大功于周室,死后获谥。这是谥法之始。《周礼》说:“小丧赐谥。”小丧,指死后一段时间。《逸周书·谥法解》:“谥者,行之迹也;号者,表之功也;车服者,位之章也。是以大行受大名,细行受细名。行出于己,名生于人。”西周的谥法制度规定:谥号要符合死者的为人;谥号在死后由别人评定并授予。君主的谥号由礼官确定,由即位的下任皇帝宣布。谥法规定了若干个有固定涵义的字,大致分为三类:1. 褒义类,称为上谥、美谥的有:文、武、明、睿、景、康、庄、宣、懿、烈、昭、穆字等;2. 贬义类,称为下谥、恶谥的有:炀、厉、灵字等,如周厉王杀戮无辜,专制残暴,遂有“厉”的谥号;3. 表达同情类,称为中谥的有:哀、怀、隐、悼等,如楚怀王的“怀”表示“慈仁短折”。1926年6月,清朝著名学者王国维自沉身亡,溥仪诏谥“忠悫”,墓碑上刻着“王忠悫公”。悫,诚实。这也是中国历史上的最后一个谥号。

鲁隐公的“隐”意味着藏匿,不显露。《史记·鲁周公世家》记载:“四十六年,惠公卒,长庶子息,摄当国,行君事,是为隐公。初,惠公适夫人无子,公贱妾声子生子息。息长,为娶于宋,宋女至而好,惠公夺而自妻之。生子允。登宋女为夫人,以允为太子。及惠公卒,为允少故,鲁人共令息摄政,不言即位。”周朝的鲁隐公是桓公的兄长,父惠王死前立桓公为太子,惠公卒,桓公“年亦甚幼小,不能为君,是隐公行国君之政,而实奉桓公为君。正所谓‘是以隐公立而奉之’”。这种特殊的“摄政”身份,使得鲁隐公成为了一位“幕后”的君主。周朝礼制完备而严格，是“周人为政之精髓”。鉴于摄政身份，鲁隐公一生在许多须行国君之礼的场合，一贯表现出谨守法度、不逾规矩的低调形象。正是由于鲁隐公一生的谨小慎微，后人赐他谥号“鲁隐公”。

《鲁隐公观鱼》的砖雕是根据《左传·隐公五年》一书中的《臧僖伯谏观鱼》而创作的。《左传》是我国古代第一部叙事详细的编年史著作。《臧僖伯谏观鱼》是其中的代表作，详细记述了鲁隐公任帝第五年春，臧僖伯谏鲁隐公如棠（鲁国地名，今山东省鱼台县东，近宋鲁边界）观鱼的过程，并对隐公“如棠观鱼”这种越礼行为提出了批评。按当时的礼法，打鱼是贱业。隐公身为一国之君，要做“大事”，如祭祀、发展军事、治理国家等。国君的言行必须要受限于礼制，要隐藏起个性，放弃自由，尽国君之责，做礼的表率。“不轨不物谓之乱政。乱政亟行，所以败也”①。身为国君去观鱼看热闹，“非礼也”。但臧僖伯劝阻无效，鲁隐公“遂往，陈鱼而观之。僖伯称疾，不从。书曰‘公矢鱼于棠’，非礼也，且言远地也。”鲁隐公不听劝阻、耽于享乐的性格与周礼相违背，与个人身份不符，是越礼行为。“犹秉周礼”的鲁隐公为什么不听劝阻，非要一意孤行呢？宋人吕祖谦的话道破玄机：“游宴之逸，人君之所乐也；谏诤之直，人君之所不乐也。以其所不乐，而欲夺其所乐，此人臣之进见，所以每患其难人也。②”游玩宴会为人之所乐，无人例外，君王亦是如此。“久在樊笼里，复得返自然。”僖伯的谏言，恰恰就是“以其所不乐，而欲夺其所乐”。鲁隐公耽于逸乐，想游离于自然山水之间而追求一时的精神逍遥与自由解脱，观鱼的行为反映了他作为人的“本真”性格。虽然鲁隐公的生活时代距老庄学说的出现早一百多年，但是这位儒家文化发源地的君主竟然做了有悖周礼、却蕴含着道家生活情趣的越礼之举，而被后世津津乐道。其实，儒道两家的思想并不矛盾，它们都揭示了人的所思所为，探索了生命的原理。

六、“祥云瑞兽”艺术语言的审美文化寓意

“中国艺术的特色是由中国文化的特色所决定的。”中国文化宇宙观认为“气化流行，衍生万物，气凝成物，气散而物亡，复归于太虚之气。天上的日月星辰，地上的山河草木、飞禽走兽和人类，皆由气生。气是宇宙的根本，也是具体事

① 左丘明. 臧僖伯谏观鱼论[J]. 读写天地. 2011(5). 20.
② 皮亮. 臧僖伯谏观鱼论[J]. 传奇传记文学选刊. 2010(5). 38.

物的根本,因而也就理所当然地是艺术作品的根本。[①]"气韵生动是中华艺术的根本精神,因为艺术描述的对象是由气生成的宇宙万物。世界上各个民族都有崇拜的动物和喜爱的植物,并且都会对它们倾注文化情感,将它们融入民族的追求和信念,以表达本民族的愿望诉求。

在花戏楼砖雕组图中有许多瑞兽的单独造型或它们与草木花卉构成的寓意图案,这些都代表着中华民族的喜好和追求。中华民族是一个非常看重宗族和血缘关系的民族。从周朝的宗法和礼制建立以来,宗法制度作为政权、族权和神权相结合的产物,就一直毫不动摇地主宰着中国的社会秩序和政治制度。中国人以子孙满堂为荣,崇尚几代同堂的大家庭,认为这样有"福气"。英文词组 extended family 最能表达中国人的愿望,即宗族血缘不断延续而形成的大家庭。中国大家庭里一般三代、四代同堂较为多见,而"五世其昌"少见,"九世同居"更为罕见。大关帝庙正门的砖雕作品,《九狮(世)同居》与《五狮(世)其昌》,便是利用汉语谐音字"狮"与"世",采用比拟的方式:砖雕五狮象征着五世同昌;九狮则表示九世同居。砖雕图《五世其昌》和《九世同居》表达出中国人对多子多孙的幸福美满大家庭的祈盼和向往,体现了中国文化根深蒂固的宗族观念。

狮子原本性情凶猛,是同类竞争最激烈的猫科动物,也是动物界顶级的掠食者。而中华文化却赋予了小狮子活泼可爱、大狮子威武高大的形象。如砖雕《狻猊》、《一品朝》。狮子又名狻猊,被佛教尊为护法兽;而狮子另一个带有政治色彩的名字是"一品朝"。在中国文化中,"'品'指古代的官吏等级,分为从一品到九品,一品最高。而狮子为兽中之王,群兽敬畏,故名'一品当朝'"。

《六合同春》(鹿鹤同春)与《三阳开泰》的动植物雕作也是借助同音字"六"与"鹿"、"合"与"鹤"、"阳"与"羊"、"泰"与"太"形成谐音比拟,以物喻人。鹿鹤同春比喻大地一片春光,生机盎然,人类共享幸福与温暖。中国的历法以正月为阳;泰指"天地交泰",冬去春来的意思,故以"三阳开泰"指岁首。旧有对联"三阳开泰,五世同昌",表达了对阳春三月,人丁兴旺、天下太平的美好向往。中国人对于四季循环变化的敏感性和对良好自然环境的祈望来自于以农业为生存之道的农耕文化。《周易》曰:"不耕获,未富也。"中国是个农业国,作为一个农业民族,中国人尚农重农,安土乐天。几千年来,中国人的主体——农民,"日

① 张岱年,方克立. 中国文化概论[M]. 北京:北京师范大学出版社,2013. 191,272,289.

出而作,日入而息。凿井而饮,躬耕田畴”,世世代代,年复一年地从事农业生产,农业生产的节奏早已与国民生活的节奏相通。根据农事农时规律,农业活动从春播、夏种到秋收,根据四季变化周而复始,“寒往则暑来,暑往则寒来”(《易传》)。在科学技术不发达、自然灾害时有发生的情况下,中国人靠天吃饭,因而他们祭天拜地,祈盼着老天爷能够“睁开眼”、“发善心”,保得年年风调雨顺,岁岁平平安安,家家都有好收成。农谚说:“一年之计在于春。”因此中国人非常重视一年之首,即农事开始的季节。正所谓春种秋收,春华秋实。“好的开始是成功的一半”,好的开春也预示着秋天的好收成。所以,这些表现“人与自然的和谐相处”的期盼的祝福语——六合同春、三阳开泰、万象更新妇孺皆知。它们都是用祥瑞的动物形象及寓意表达中国人对大地同春、风调雨顺、太平盛世和美好生活的盼望。

花戏楼正门和钟楼、鼓楼门头上的纹饰小品《二龙戏珠》和《龙凤呈祥》是采用简化的手法雕刻的作品。砖雕装饰艺术与文字和绘画艺术不同,因受构件材料和面积的限制,往往需要用简化的形式,概括性或象征性地表现作品的思想内容和文化内涵。龙和凤都是中国文化创造出来的神物,是人们通过想象创造出的动物,是中国人崇拜的图腾。在中国人的生活和艺术创作中,龙、凤、麟、龟作为祥瑞动物受到推崇。《礼记·礼运》载“麟、凤、龟、龙,谓之四灵”。“四灵之中,除龟为实有动物外,龙、凤、麟都是由远古图腾演变而来的理想神物。它们的历史久远,所象征的民族意义非常深刻,反映出中华民族追求和平幸福、吉祥长寿的文化心理。[①]”

麟,就是麒麟。早在《史记索隐》中就有记载:“雄曰麒,雌曰麟,其状麇身,牛尾,狼蹄,一角。”而《毛诗义疏》曾曰:“麟身马足牛尾,黄色,圆蹄,一角,角端有肉。”但一般仍然认为麒麟是人们通过想象将某些动物的特征组合起来塑造而成的,“它是由鹿及其同类演变而来的”。它极有可能是周族的图腾。古人认为鹿是“纯善之兽”,是纯洁善良和道义的象征,故有“道备则白鹿见”一说。因此麒麟象征太平统一,也比喻杰出人物,故称聪明伶俐的小孩儿为麒麟儿。

“凤凰的形象实际上来源于古石器时代的鸵鸟”,传说中的鸵鸟能辨听音乐,善于舞蹈。凤凰被看作具有仁义道德的女性形象的象征。

---

① 程裕祯. 中国文化要略[M]. 北京:外语与教学研究出版社,2011. 383.

从远古时代,龙就与帝王联系在一起。自汉高祖开始,中国历代帝王为稳定自己至高无上的地位,自喻为"真龙天子",确立了帝王"九五之尊"的至高无上的地位。在民间,龙是奋进和力量的象征。按照中国"一阴一阳是为道"的宇宙观,自然界中的天地、阴阳、乾坤相合,而龙与凤一雌一雄也代表阴阳相合,因此龙凤是古代建筑装饰中很常见的艺术形象,也是中国文化中最能代表喜庆吉祥的祥瑞动物形象。

花戏楼鼓楼右下方的兜肚《喜鹊登梅》砖雕刻有一对喜鹊落在一株怒放的梅花枝头,梅花仿佛暗香浮动,两只鸟儿一鸣一啄,十分有趣。梅是百花中开在一年之首的"报春花",故有"春为一岁首,梅占百花魁"的春联。喜鹊叫声婉转,有"报喜鸟"的别称。在中国民间人们认为听到喜鹊在枝头的叫声预示着家中喜事将至。喜鹊登上梅梢(梅花枝头),构成谐音比拟"喜上眉梢",极具喜庆之意。

中国文化强调"天人合一"的思想,重视人与自然的和谐统一。因此,追求"和"是中国艺术的最高境界。在这种理念的支配下,自然界的一切都被艺术赋予了生命。在中国文人的笔下,梅、兰、竹、菊因其清雅淡泊被称为"花中四君子",松、梅、竹因其耐寒长寿被称为"岁寒三友",它们都寓意着人的品格高尚。从古至今,赞颂梅花的诗词层出不穷。毛泽东笔下的梅花品格中有乐观超然的成分:"风雨送春归,飞雪迎春到。已是悬崖百丈冰,犹有花枝俏。俏也不争春,只把春来报。待到山花烂漫时,她在丛中笑。"陆游赞颂梅花的节操,却也表现出它的悲凉孤傲:"驿外断桥边,寂寞开无主。已是黄昏独自愁,更著风和雨。无意苦争春,一任群芳妒。落成泥碾作尘,只有香如故。"

钟鼓楼砖雕作品中的《松鹤延年》、《福寿图》、《寿比南山》和寓意幸福美满的《凤凰戏牡丹》和《鸳鸯戏莲》等,运用动物和植物寓意延年益寿、团圆和美、幸福美满。松柏往往在寒冷中显出与众不同的品格。《论语》曰:"岁寒然后知松柏之后凋也。"《庄子》云:"天寒既至,霜降既降,吾是知松柏之茂。"古人把松柏看作"百木之长",能抗严寒而常青。鹤在中国文化里是长寿鸟,汉语里称老人的年龄为"鹤龄"。故松鹤都寓意着长命百岁、延年益寿的期望。牡丹素称花王,以其雍容华贵的特点深得中国人的喜爱。它象征荣华富贵、幸福美满。莲在周敦颐的笔下是"出淤泥而不染"的花中君子。莲又名"水芙蓉",比喻纯洁美丽的女子。另外花戏楼正门上的砖雕图《犀牛望月》、《万象更新》都是借用动物形象寓意人类对美好生活的期望,象征吉祥如意、喜庆欢乐,表现出中国文化含蓄深沉、意味隽永的艺术审美追求和“中和之美”的文化特征。

出于明清时期的商帮行会经商、祭祀、娱乐等活动的需要，全国各地兴建会馆无数，山陕会馆只是其中一类。如今时过境迁，许多会馆已难觅踪影，唯独山陕会馆留存下来的数量最多，也较为完整。山陕会馆历经千百年的战乱烽火、风吹雨打，能保留至今，不能不说是个奇迹。究其原因，大致如下：首先，晋商、秦商两大商帮清朝时资金基础雄厚，帮派内部团结，重乡情、尚义举；他们强强联合建馆，资金较为充裕。被誉为“天下第一馆”的河南社旗会馆《创建春秋楼碑记》记载，修建大拜殿（关公祠）及其附属工程的花费高达 88788 两白银。社旗会馆现存的《公议杂货行规碑记》载：“据考证，在乾隆年间，三两白银可供八口之家一年的口粮。”相比之下，八万八千多两白银可谓是天文数字。耗资之巨大，非富不可为。其二，建馆所用材料坚固，建筑人员都是来自各地的能工巧匠，因此建筑质量较高。在许多山陕会馆的兴建过程中，建造者往往是“运巨材于楚北，访名匠于天下”。社旗、聊城和亳州山陕会馆、徽州会馆、江宁会馆、福建会馆等都是如此。明清时期，这些商贸要地四通八达，会馆林立。如今，江宁会馆残破不全，其他会馆只剩断壁残垣，或早已化为灰烬，只剩下山陕会馆依然挺立。其三，山陕会馆都建有祭拜关帝的大关帝庙，并以其为主要建筑。关帝是儒释道三家的崇拜对象，在中国文化中地位很高，在国人心里神圣不可践踏。所以关帝庙可能会得到当地民众更多的维护和关注。其四，晋商的诚实守信、仗义厚道、帮派团结等特点可能会赢得更多民众的信任和好感，所以民众对晋商建的会馆较为爱护，而不会轻易加以破坏。第五，山陕会馆的建筑结构往往比较合理，都是两进、三进，甚至五进的封闭型四合院，适于改建后为政府使用或用作居住。如山东的聊城会馆，因其与聊城一中隔河相望，“文革”期间被用作学生宿舍而得以幸存下来。会馆内祭拜的关帝是武将，虽非“封建老祖孔子”，也曾被认为是“祭拜鬼神”的“封资修”。在“文革”期间“破四旧”中，造反派和红卫兵小将们对关帝庙进行了“打砸抢”。由于一些文物专家和爱心人士的巧妙保护，许多会馆才幸免于难。据说当时一位老教师为了保护社旗会馆“大拜殿”牌坊的中坊下入口处的巨大青石雕“九龙口”，巧妙地利用了建筑材料做成一个“忠”字覆盖于上，才使其得以留存。而为了保护亳州花戏楼砖雕，人们当时在它们的表面涂了泥，上方覆盖红色标语，才使其幸免于难。亳州花戏楼砖雕历经几百年能保留至今而几乎完好无损，仿佛冥冥之中有神明庇佑，让我们得以欣赏到如此精致的艺术瑰宝。

花戏楼砖雕是不可再生的精美艺术品,它给我们留下了一笔巨大的精神财富。在花戏楼山门上的整个砖雕序列作品中,无论是表现和平还是战争的场景,是描绘贫穷还是富贵的内容,无论使用何种手法,没有一幅作品表现出悲观厌世的情绪,或是宣扬消极颓废的思想精神:从反映高官生活的《郭子仪上寿》,到民间传说《李娘娘住寒窑》;从充满硝烟的《吴越之战》到悠然自得的《陶渊明爱菊》。在所有的作品中,连鸟兽和花草也被寄予了美好的情感,如《松鹤延年》、《喜鹊登梅》、《鸳鸯戏莲》等,都传达着美好的祝福,呈现出祥和的文化氛围,没有一丝悲凉凄切或消极颓废的感觉。它反映出处于上升态势的晋商的文化心理和精神状态,它宣扬的是一种积极向上的、追求善与美的思想精神和生活态度。花戏楼砖雕用特殊的“图视”艺术语言符号传达给观者和后人的是中国传统文化“积极有为,奋发向上”的精神和“生生不息”的生命哲学。这就是花戏楼砖雕艺术作品所具有的宝贵的中华文化特质,也是经久弥新的砖雕艺术的魅力之所在。

第二部分

The Cultural Aesthetic Implication of the Brick Carvings of Bozhou Gorgeous Dramatic Stage

Chapter 1

The Art of the Brick Carvings of Bozhou Gorgeous Dramatic Stage

An ancient theatre named Gorgeous Dramatic Stage (Huaxilou in Chinese) was built over 300 years ago in Bozhou City, Anhui Province. And now it still stands as it used to be, world-renowned for its carvings and colored pattern crafts, especially the exquisite brick carvings.

Constructed with the contribution of businessmen from Shanxi and Shaanxi who lived in Bozhou in the Ming and Qing Dynasties, the Gorgeous Dramatic Stage is also named Shaanxi and Shanxi Guild Hall. Its use for worship of Guan Yu earned its original name "The Grand Temple of Guan Yu". It is a temple with a dramatic stage in it, integrating the function of holding religion rituals and business parties as well as entertainment. Its another name *Huaxilou* in Chinese owes to the gorgeous dramatic stage decorated with beautiful paintings of flowers and exquisite brick carvings. It is now considered as a place of historical interest and a scenic spot in Bozhou, a first class unit of national cultural relic preservation. Nowadays, the group of buildings, named the Huaxilou Scenic Spots, dominated by the Grand Temple of Guan Yu and the dramatic stage makes it a must-go place for visitors when they travel to Bozhou.

Since Jin in Chinese is the abbreviation of Shanxi Province, the builders are called Jin Merchants, that is, Shanxi Merchants. Bozhou Gorgeous Dramatic Stage served as a liaison place for them to do Chinese herb trade, simultaneously, functioned as a center of commerce, entertainment, religion and politics. For more than three hundred years, the Gorgeous Dramatic Stage witnessed the transition of times and society, as well as the history of Jin merchants' culture. Huaxilou survived and then flourished, with an exquisitely decorated dramatic building swarming with striking carvings left. For the local people in Bozhou, it is more than a dramatic building or a stage with fine carvings; it represents a long sculptured history, a period of past time, flourishing recollections and extensive culture.

Ⅰ. The Three Wonders of Bozhou Gorgeous Dramatic Stage

More than 80 ancient building groups served as guild halls remain there in China. Among them there are four Shanxi and Shaanxi Guild Halls in Henan Province, specifically, in Sheqi, Tanghe, Dengzhou and Luo cities; one in Liaocheng of Shandong Province and by the Jinghang Channel; one in Xiangyang, Hubei Province, etc. Their construction is a vivid expression of architectural achievements in imperial courts, temples, guild halls, citizen buildings and gardens. They are all full of magnificent and exquisite carvings. They represent the momentum of imperial palaces. They abound in soft and beautiful colors and qualities of poetry and painting, which leave us great impression of overall beauty. It is taken as the masterpiece of the ancient Chinese structures in distribution, overall design and decoration art. Bozhou Huaxilou (Bozhou Shanxi and Shaanxi Guild Hall) and Sheqi Shanxi and Shaanxi Guild Hall (recognized as "Number One of guild halls in China" by relic experts) were designated as national key spots of cultural relic preservation in January, 1988, by the Chinese government. Though Bozhou Gorgeous Dramatic Stage is smaller than Sheqi Guild Hall, it predates the later according to written history.

Huaxilou Scenic Spots composes of the Grand Temple of Guan Yu, the dramatic stage, Temple of Zhang Fei, Zhu Gong Academy of Classical Learning, Temple of God of Fire, cereal lane and the guild hall. The Grand Temple of Guan Yu facing south is the main part which is a three-storied wood-like structure with exquisite brick carvings. It is flanked by Bell Tower and Drum Tower and its main gate is on the ground floor of the nationally famous Huaxilou which is closely adjacent and behind it. When going further through the gate, a court presents itself in front of you; on its both sides there are watching buildings. At the north end of the court is a hall which consists of front and back parts. The whole building is full of exquisite sculptures and other delicate decorations, with an area of only 3163 square meters.

In Chinese history, several thousands of years leave later generations innumerable glorious imperial palaces, brilliant temples, beautiful gardens and other kinds of houses of different shapes. It's safe to say that decoration plays an important role in these buildings. On the side of the drama building, the cubic terrazzo brick carvings embedded on the main gate of the memorial archway structures are beautiful, incomparable and famous. The brick carvings representing consummate artistic carving, exquisite skills, and ingenuous delicacy are distributed carefully and neatly and recognized as a masterpiece, taking up a key place in the national decoration art. Whoever goes to Huaxilou and sees those gorgeous brick carvings will be astonished by the innumerable amazing stories brought about by the little carving.

In a sense, Shanxi Merchants and other men of eminence regard the halls as show-off signs of wealth which equal modern advertisements. Therefore, they are more than willing to invest lots of money for building and decorating guild halls to make them brilliant. Thus Bozhou Shanxi Guild Hall presents gorgeous art of building, decoration and well-examined technology. Of all the artistic workings in the Gorgeous Dramatic Stage, the brick carvings, iron flag post and painted wood sculptures are recognized as its three wonders.

The main gate, a wood-like memorial gateway facing south replete with many brick carvings, is shared by Huaxihou and the Grand Temple of Guan Yu. It is flanked by the arched gates of Bell Tower and Drum Tower. The memorial gateway is decorated entirely by water-milled brick carvings which are less than 10 centimeters thick with exquisite figures, mountains, chariot, horses, city wall and moat, insects, fishes, birds, beasts, trees, flowers, grass, kiosks and buildings in them. Because of the life-like figures, wonderful buildings and exuberant trees and grass in the carving, it is acclaimed as one of the three wonders of the Dramatic Stage.

The second wonder is the two iron flag posts in front of the main gate of Huaxilou. They are 16 meters high and weigh 6 tons each, around which are winding iron cast dragons with their heads and tails raised, lovely and vivid. On both of them there are three layers of square dippers with eroded bells. They will ring when wind comes, with the sound clear and melodious. The iron flag posts surprise the visitors with its height, magnificence, unique modeling, solidity and endurance.

The third wonder is the brightly painted wood sculpture, the soul of Huaxilou. Many of them are on the sides of the drama stage telling the stories of the Three Kingdoms Period. Not only the figures in the wood sculptures are vivid, but the city walls and moats in it also give a sense of three-dimension and authenticity. The wood sculptures of the dramatic building are completely painted with deep, bright and obviously distinctive colors. While wood sculptures and paintings are combined ingeniously together, they appear to be complementarily vivacious. The brick carvings, the iron flag posts and the painted wood sculpture are of different characters and are recognized as the three wonders and essence of Huaxilou, as three bright pearls.

It deserves to be studied that whether the world surprising brick and wood sculptures of Huaxilou supposedly belonged to the style of Hui School structure in the past and the contribution was attributed to Hui businessmen. It was later examined and confirmed by an expert in ancient structures, Professor Chen Congzhou from Shanghai Jiaotong University, that Bozhou Shanxi and Shaanxi Guild Hall was built by craftsmen from Shanxi Province, belonging to the style of Shanxi School structure. If his conclusion is generally accepted, the style of Shanxi School structure art will have a new area of study. But some younger experts put forward their opposite views that the constructive style has followed that of Hui School in the fine design and the elaborate skill of brick carvings.

Ⅱ. The Decorating Art of Brick Carvings

1. A Brief Introduction of Chinese Brick Carvings

The decoration art of brick carving is a unique kind of decoration art in traditional Chinese structure, together with wood and stone carvings. In ancient China, brick carving is widely used as civilian house decoration. Folk craftsmen use chisel and wooden hammer to carve pictures and decorative patterns with specially made bricks by sawing, drilling, carving, chiseling and polishing. In this way, the bricks are processed into decorative artworks of different figures, flowers, birds and beasts and auspicious designs and then placed on the arches, the ridges of roofs, corner belts, gables, shadow walls, upturned eaves and railings, and on the wall surfaces, etc. They function as decoration of buildings, embodying the temperament and interest of the designers as well as their appreciation of beauty and wishes.

The raw materials of brick carving are of easy availability, high wearability, and anti-corrosion, so brick sculptures are widely used for official and civilian buildings and temples from the Song Dynasty. Brick carving patterns are of abundant contents and extensive themes, presenting figure, mountains, water, flowers, birds, beasts, carriages, horses, city walls, moats, geometrical objects and characters, etc. Figures dominated themes are of religion, myths, drama, folk customs and social life, etc. Brick carvings are of consummate skills, freshness, simplicity and wonderful workmanship beyond this world. Its different forms of art, plentiful expression contents and deep cultural intention show the characters of national and regional cultures. The unique cultural connation and wonderful melted-down structural decoration art of Chinese brick carving made it to the list of national level non-material cultural heritage and gained its fame in the gallery of international structural art.

The art of ancient Chinese structural sculpture and handcraft articles of gray brick carving were the evolution of the tile art in the Dongzhou Dynasty and the painting bricks in the Han Dynasty. It is one of the important art forms to carve mountains, water, flowers, figures, etc on the gray bricks used for decovation. Brick carvings usually refer to gray brick carved handcraft articles. With its appearance in the Han Dynasty, development in the Song Dynasty and heydays in the Qing Dynasty, the brick carvings' development is closely related to structural materials. China is a land dominated country, so ancient buildings were mostly of civil-engineering structure. Different from stone buildings in ancient western countries, advanced buildings in ancient China were mainly made of brick and wood, and with civilian ones, soil, wood and grass. Brick evolved in the development of pottery. Our ancients learned to carve decorative earthen patterns on the surface of pottery. They learned the decorative skills of carving, molding, printing, attaching, etc. As early as in the new Stone Age more than 7000 years ago, the ancient Chinese people mastered the skill of earthen baking. In the about 4000-year old Siba cultural relics, people found finished sun-dried bricks which were made of soil and other materials. During the Warring States Era, the transition from earthen brick to baked brick was realized. It was initially used for decoration and widely used in grave structures. Bricks made in the Qin Dynasty enjoy a good reputation, so there was an idiom "Bricks of the Qin Dynasty and tile of the Han Dynasty", meaning bricks and tile made at those times are of good quality. Bricks of the Qin Dynasty were mainly earthed in the tomb of the first emperor of the Qin Dynasty. They are grey, solid, durable, regular and massive. Especially, the diversity of mold printed decorative patterns is famous far and wide for the sense of plentiful gradations produced by veins of close texture.

In the West Han Dynasty, painting bricks were more of decoration, with abundant themes. The unfired painting bricks were streamlined and molded, so the same content and images appeared again and again. The unearthed painting bricks of the East Han Dynasty have represented an independent and complete artistic work on the brick with an artistic conception of diversity. And the painting brick of tomb in the Han Dynasty was hollow and massive, manufactured by pressing painting molds onto the wet unfired bricks which evolved into brick sculptures for decorating the walls of coffin chamber in the North Song Dynasty. The three walls of excavated coffin chamber of the North Song Dynasty were covered with brick sculptures in the provinces of Henan, Shanxi and Gansu. The quantity, quality and themes of the brick sculptures in the coffin chamber were decided by the social position of the tomb owners. It is commonly seen in the sculptures that the owners of the tomb sit opposite each other, with menservants holding a tray and servant girls holding a kettle, which is a reappearance of the tomb owners' life before their deaths. In the Jin Dynasty, the contents of brick carvings of coffin chamber were more diversed. Brick carvings of coffin chamber were still an extension of the Wei and Jin Dynasties. During the period of the Jin and Yuan Dynasties, the tomb brick carvings developed into a larger scale. Dong Feijian tomb, built in Houma, Shanxi Province in 1210, with a coffin chamber area of less than 4. 7 square meters, is alive with brick carvings which include wood-like structures: *dougong* (a system of brackets inserted between the top of a comb and a crossbeam), *gongyan* (the space between *dougong*), sunk panel, gate, partition board, etc and paintings of screen, stools, flowers, birds, beasts, figures, acting scenes, etc. The characters of *sheng*, *dan*, *jing*, *mo* and *chou* in Beijing opera, carved three-dimensionally are lifelike, which is the masterpiece of Jin Dynasty brick carvings.

In the Yuan Dynasty, brick carvings of coffin chamber were declining gradually. Shanxi Province was a booming place of Chinese dramas. A lot of drama themed brick carvings were found in the graves. In the Ming Dynasty, it was widely used as building materials in the folk residences and palaces, and the role of drama-themed brick carvings changed from grave ornament to regular architectural ornament. It was an important ornament form of the academies of classical learning, guild halls, ancestral halls, temples and theaters, etc.

In the Qing Dynasty, with the support of commerce, it reached the peak. The interregional cultures also spread and fused. The earliest salt and Chinese herb merchants from Shanxi and Shaanxi Provinces brought them to the valleys of Yangtze and Huai River as well. For instance, Suzhou Municipal Drama Museum used to be Three Jins Guild Hall, where drama characters brick carvings are seen above its gate and on its arch over the gateway. The same carvings appear in Tianshui Shanxi Guild Hall in Gansu Province as well. The brick carvings named *The Whole Family Picture of Guo Ziyi's Birthday Celebration* and others have made a name for Bozhou Huaxilou. The air vents of the pillars attached to the inner walls make good use of the brick carvings. The penetrated carvings with flowers and birds were solid, beautiful and beneficial to the airflow. The brick carvings were also attached to the walls of emperor's resting place and its companion structures owned by Cixi, the Empress Dowager, some of which were covered with splendid gold leaves.

In material selection, the soil for unfired bricks had to be tiny, for it was first screened, washed and faded carefully. After many manufacturing processes, it was made into refined unfired brick, then fired into fine, smooth and regular gray bricks. After washed and rubbed, they were carved carefully by using different carving skills according to the themes. The inanimate soil was as if endowed with magic and became fine brick carvings.

The characteristics of Chinese brick sculpture art are mainly used for decorating architectural components of the main gate, shadow wall and wall surface. Strict requirements for material selection, forming, firing and so on lead to the solidity and fineness of bricks appropriate for sculpture. The brick carvings provide a near and far view with a complete artistic effect. It is thematically dominated by mainly *Dragon and Phoenix Indicating Good Fortune*, *Liu Hai Plays with the Gold Toad*, *Three Yangs Bring Auspices*, *Kylin Sends Kids*, *Pine and Crane of Longevity*. The plants and flowers like pine, orchid, bamboo, camellia, chrysanthemum, lotus, carp, etc., are all with propitious implications, enjoying popularity among the people.

There are many ways of producing brick carvings, such as deep carving, spherical carving, carving by outline drawing, shallow relief sculpture, openwork carving, and flat carving. From the angle of practicality and ornament, brick carvings for civil use are terse in image, simple in style and deep in meaning. It can maintain solid and stand the erosion of sunlight and rain.

2. The Brick Carvings of Hui-style in South China and Jin-style in North China

Chinese brick carvings enjoy not only a long history but also an integration of the regional and ethnical culture, presenting both fusion and differences in style and artistic characters. Generally recognized brick sculpture schools in China are as follows: 1. Beijing style; 2. Tianjin style; 3. Shanxi style; 4. Huizhou style; 5. Suzhou style; 6. Guangdong style; 7. Linxia style.

It is generally agreed that the best developed Huizhou brick carvings and its civil brick sculpture evolved in Yangzhou and Hangzhou are ahead of other schools to the south of Yangtze River. Huizhou brick carvings (Huizhou refers to current Xi County in Anhui Province) stemmed from the Song Dynasty. The Ming Dynasty favored primitive simplicity, relief carvings and single gradation. The Qing Dynasty was fond of multi-gradation open work carving, with water mill bricks as material and superb skills. It reached its heyday in the Qing Dynasty. There were only flat carving and shallow relief sculpture emphasizing symmetry, full of decorative delight with the aid of lines and modeling, but perspective changes were to be desired. Its composition and distribution of pictures adopt the expression means of Xi'an painting school, which favors aesthetic beauty, with frequent use of deep relief carving, spherical carving and gradations, with sometimes more than 10 layers on one brick. The expanding hollow-out represents a collective picture of kiosks, buildings, mountains, water, trees, figures, beasts, flowers, insects and fishes, looking like a wash painting. Its unique beauty earns itself national and international fame. The brick carvings *The Whole Family Picture of Guo Ziyi's Birthday Celebration* in Bozhou and Anhui museum are representatives of Huizhou brick carving, with excellent art style and consummate skills.

2.1 Huizhou Brick Carvings

In southern Anhui Province, civil residence with white walls and black tiles decorates the green fields and lush mountains. Grey roof with door ornament and upturned eaves is the characteristic of Huizhou buildings, in harmony with green mountains and limpid water nearby. The style of Huizhou brick carvings gives primacy to plot and composition of pictures, characterized by multi-layer penetrated carvings. Brick carvings in Huizhou are elaborate, with neat carving, smooth lines, outstanding themes and clear gradations. Pictures of complex plot and multiple layers are carved on an unfired brick of more than one *chi* (a unit of length) square big, less than one inch thick. The distant and near scenes can be distinguished with clear gradations. A square brick can be carved into nine layers at most. The distribution of the whole pictures is in the way of vertical scroll or hand scroll. It creates an atmosphere of sedateness, offering an incomparable sight. Its carving skills are brought to an art form.

The verse "Door ornaments are as beautiful as temple ornaments, and shadow walls become sculptured walls" is a real portrayal of Huizhou brick sculpture. Huizhou brick sculpture gives primacy to the ornament of the arch over the gateway and the door. The gateway and door ornaments, as a badge of entry and exit, are of great variety. The door ornaments of Hui-style civil buildings, protruding line angle ornaments built with water wash bricks and covered with tiles, are on the outside frame of the main door. For example, the gateway of the ancestral hall of Pan family is of five phoenixes style, providing a grand sight. The walls in the shape of Chinese character "八" on both sides of the gate hall are decorated with many fine brick carvings of beautiful scenes to the south of Yangtze River, buildings, kiosks, waterside pavilions, birds, and beasts. The use of spherical sculpture and proper rise and fall of the sculptures lend them unique lingering charm. They look like a wash painting, pure, fresh, simple and elegant. The crossbar linking two columns, especially the panoramic crossbar is the most wonderful part of the door ornaments of the arch over the gateway of Huizhou civil buildings. The use of shallow and high relief, penetrated or semi-spherical ones or hollow-out ones bring a sense of distance and gradation. The variety of brick sculptures on the ridge and its both ends can be seen on many temples, ancestral halls, and civil buildings.

The content of Huizhou brick carvings is wide-ranging, from flowers, figures, drama, auspicious decorative patterns to life scenes. The figure dominated themes include historical stories, drama, folklore, religious mythology, legends and people's social life, such as the life portraits of kings, generals, prime ministers and aristocrats, journey of businessmen, life of scholars, woodcutters, peasants, shepherds, as well as performances of recreation, such as lantern show, dragon dance and lion dance. For example, *The Whole Family Picture of Guo Ziyi's Birthday Celebration*, *Eight Immortals Crossing the Sea*, *Picture of a Hundred Children*, and so on. Animal, flower, insect and fish themed brick sculptures implying luck are mainly found on the arches over the gateway, door shelters and junctions of cross beams and posts, such as, *Dragon and Phoenix Indicating Good Fortune*, *Peony and Phoenix Facing the Sun*, and so on. The painting of flowers are various and colorful. Pine, bamboo and plum, peony and so on are often central to the carving theme. These images are created by special skills.

Implying propitiousness, joy and other good wishes, each picture is meaningful. The brick carvings to the south of Yangtze River evolve under the influence of Huizhou brick sculpture. It looks more refined, skillful, fresh and smooth, with neat sculpture skills. Brick carvings in Lingnan civil residences are technically freer and thematically richer. It is especially typified by the three large brick sculptures gazing at the visitors in Chen's memorial temple in Guangzhou city. *Liu Qing Tames the Horse Langju* on the east wall has 30 plus figures, *Liangshanbo Heroes* on the west wall presents over 10 figures, all with different postures, clothes and expressions sculptured clearly and carefully. Their use of relief, penetrated and cubic sculptures presenting the relation between the figures and the situation makes them coexist in a space of many gradations in an orderly way. Guangdong brick sculptures are often carved with hollow-out skills, usually called Silk Sculptures (挂线砖雕) with delicacy and ingenuity, against the roughness and simplicity of the ones in the north. The mix of concave, shallow relief, high relief and penetration sculptures is used. The refined sculpture has as many as 9 gradations, producing a visual effect of both near and distant scene. Their colors will range from black, white and grey because of different times in the day.

2.2 Brick Carvings in North China

The brick carvings in central Shanxi Province, Beijing and Tianjin are of mature technique, terse modeling, simple style, solemnity, boldness and vigor. The brick sculpture in Shanxi Province is durable because the unfired brick is made of quality soil. The painting and carving skills are refined and unique, with clear pedigree. We have famous brick sculptures on the walls of Qiao and Chang's courtyard in Shanxi Province, the Summer Palace and the Imperial Palace. Brick sculptures of air holes and shadow walls will be handed on from age to age. They pay equal attention to the ornaments of such arch over a gateway, but wall, railing and shadow wall brick sculpture are more commonly seen in the north, comparing to that of the south. Their carving skills are awkwardly simple, old, strong and steady, compared with Huizhou brick sculpture. The brick sculpture in Shanxi Province pursues integrity of the composition, perfect shape and the artistic effect of wholeness.

In North China, the *siheyuan* or quadrangle dwellings, a yard formed by four inward-facing houses, is a typical traditional Chinese architectural layout. "South" became a key component in Chinese culture, and south facing a *Fengshui* principle of traditional Chinese structure. China lies in the northern hemisphere, with low and mid latitudes. The geographical factor decides that the south-facing houses can get warmth facing the sun against the wind in winter and stay cool facing the wind in summer, so Chinese houses mostly face south. Geographical and natural factors cause the differences between the northern and southern Chinese structures in whatever area. Southern structures, such as temple and civil residence are usually built near the mountains, laid out according to the features of such places. Since the land in south is limited, the structural layout has to be compact. Refinement, indirectness and plentiful styles are commonly seen in the constructions in South China.

Vast territory and many plains in North China lead to the popular layout of the *siheyuan*. Plenty of space in the north makes it common for the royal and official residences to use screen wall with brick carvings as ornaments. Screen wall is an isolated wall, located outside or inside the entrance gate of the *siheyuan*, with a distance from the main gate. The screen wall outside the entrance gate is often seen before the houses, temples and residences acting as the indication of the main gate position and requirement of the passengers to avoid it. The one inside the entrance gate was used to prevent outsiders from peeping and keep concealment and quietness inside. They are called respectively *yin* (meaning to conceal) and *bi* (meaning to avoid), hence *yingbi*(影壁). No matter the screen wall is outside or inside the main gate, it would face the people who enter or leave, so it is also called *zhaobi*. A screen wall acted as an indication of wealth, position and showed the pursuit of the owner. So in North China, from emperors to the civilians, all took screen walls with fine brick carvings as an important ornament of the construction.

For example, the screen wall of the Qiao's in Shanxi Province has one hundred Chinese characters, *Shou* (longevity) on it, ten horizontal and ten vertical, forming a square. They are gold, of different forms, and on a solemn black backdrop. In addition, the screen wall of Chang's courtyard in Shanxi Province is considered to be imposing. The courtyard is horizontally wide, with six screen walls, large or small. One of them leads to the inner courtyard with two *chuihua doors* standing side by side (the inside door in the *siheyuan* is called *chuihua* because its eave columns hang in the air with painted petals), flanked by two screen walls. Between them is another set of screen wall (including a main one and two subordinate ones). There are five screen walls in all, one big, the other four small. They are of primitive simplicity and good taste. The pedestal of each is supported by *xumizuo* (莲花座,须弥座, used for supporting the figure of Buddha), standing side by side.

3. The Brick Carvings of Bozhou Gorgeous Dramatic Stage

Chinese architectural culture culminates in the Ming and Qing Dynasties. The architectural style of Bozhou Gorgeous Drama Stage represents a refined ornamental art of ancient Chinese structure at that time. The group of brick carvings is located on the gate. Besides the gate there is the colored dramatic stage.

The gate of Bozhou Gorgeous Drama Stage is also the Temple of Guan Yu, which is a wood-like brick carving memorial archway, facing the south with two symmetrical arched doors on both sides. Served as a place to worship Guan Yu, its entrance gate is also called *shanmen* (mountain gate) because temples were usually built in mountains away from secular world, so the name is given to the main gate of a temple. In Chinese, mountain and three has similar pronunciation "shan" and "san". Since *shanmen* is identical to *sanmen*, which means "three doors", symbolizing the three doors for extricating oneself in Buddhism, namely, "kongmen (the general name for Buddhism)", "wuxiangmen (entering Buddhism field through this door)" and "wuzuomen (you have no worry entering Buddhism field through this door)". The mountain gate of Gorgeous Drama Stage consists of the entrance gate and the gate of Bell Tower and Drum Tower to its east and west respectively.

The main gate is a three-tiered memorial brick arch with three-dimensional carvings. On the bricks that are less than 5 centimeters thick, 115 characters, 33 birds, 67 beasts and innumerable plants are carved, telling classic stories from Chinese religions, history, politics and customs. 52 works were carved on polished gray bricks of only about ten square meters big and less than two inches thick. The carving content is also rich, with six operas of *xiwen* (a kind of drama), 16 stories of figures, and 24 animals. The beauty is lent to Gorgeous Dramatic Stage by the carvings of Chinese characters "福" (blessing) and "寿" (longevity), and other decorative patterns, including sunflower, entangled flower and propitious cloud, etc.

The polished brick wall is full of dimensional carvings, among which the most eye catching one is *The Whole Family Picture of Guo Ziyi's Birthday Celebration*. Guo Ziyi was a meritorious statesman who suppressed the An Lushan Rebellion in the Tang Dynasty and was conferred the Prince of Fenyang (a place in Shanxi Province) by Emperor Xuanzong of Tang. The carving work displays the scene of Guo's 60th

birthday. Congratulations were offered to him by his family members, relatives and colleagues. 42 characters in total were carved with different postures, old and young, all shinning with happiness and full of joy. In the center of the hall sits Guo Ziyi, looking kind with whiskers falling down to his chest. A Chinese character "*shou*" (Longevity) can be seen clearly on the wall behind him. The civil and military officials of the imperial court stand in proper order, paying tribute or making bows, each with distinctive expression. The bustling chariots, carriage horses and citizens reveal a scene of prosperity and harmony. Being uniform and proportionate in the whole layout, the stunning work is fine and exquisite.

The wonderful stories are shown with carving art in such a small space. On the Drum Tower there carved *Three Visits to the Thatched Cottage*, which is acclaimed as the acme of carving. It is about in the Three Kingdoms era Liu Bei making three visits to Zhuge Liang's house to invite the Zhuge to join his army. On the left side there are Liu Bei, Guan Yu and Zhang Fei. Before them is Zhuge Jun, Zhuge Liang's brother, with his hand behind his back. In the story Liu Bei came to him and asked, "Is your brother in? Can we see him now?" He said, "He came back last night. You can see him." With the word, he left haughtily. On the right side of the carving is Zhuge Kongming sleeping. A poem "Who first knows my dream? No one gets its knowledge but me. I have enough spring sleep in the thatched cottage, till the sun rises so high outside the window." was created by him then. This carving work presents consummate skills, with distinct gradations and high artistic value. In the picture, a tea boy is making tea at the bedside. A teapot, a desk, a chair, a writing brush, an inkstone, a water jar and two buckets can be clearly seen. In addition, two thin trestles obliquely supporting the window shelter make the scene alive. A pair of Zhuge Liang's shoes are placed before the bed, as if they were real. Three hundred years' exposure to the sun, rain and frost causes the erosion of the stone lions, but the brick carvings keep intact, proving the perfect ancient Chinese brick firing skills.

Under the characters "钟楼" (Bell Tower), the carving named *Madam White Snake* is originated from a legend in China. The vivid figures, refined scenery, the broken bridge, flowing water, Leifeng Pagoda, temple gate and courtyard of Jinshan Temple make people feel personally on the scene and give flight to the imagination.

There are also some novel works of symmetrical layout such as *The Drawings of Four Loves*: *Wang Xizhi's Love of Geese*, *Tao Yuanming's Love of Chrysanthemum*, *Zhou Dunyi's Love of Lotus* and *Duke Yin of Lu Watching Fishing*. The carving *A Tiger Trapped on the Plain* comes from the idiom "If the tiger left the mountain, it would lose its power; if the dragon was trapped in the shallow beach, it would be in trouble". In the picture, a tiger is standing on the plain without a shred of its prestige, while two dogs are vigorously barking at it. It implies that a man who lost influence would be bullied by snobs. This sentence comes from an ancient Chinese children's primer, *A Collection of Witty Folk Sayings*. It dates back to the drama *Peony Pavillion* written in Ming Dynasty, edited by men of letters in the Ming and Qing Dynasties. It is full of folk stories with wisdom and philosophy.

There are also carvings such as *Nine Lions Playing the Ball*, meaning "may all your wishes are fulfilled" and *Five Lions Playing the Ball*. In Chinese "lion" and "generation" are of similar pronunciation, meaning "being prosperous generation after generation". Other brick carvings on the main gate, such as *Kuixing Appointing Zhuangyuan* (*the Number One Scholar*), *The Oriole Repaying the Benefactor with Jacle Bracelets*, *The Fury Toad Fighting the Lion*, and so on, all these materials are from ancient Chinese allusions. They are full of meaning and food for thought.

With rich content, clear-cut theme, free and primitive composition, uniform layout, fine cutting and exquisite carvings, the whole brick memorial archway has a strong artistic impact which is rarely seen in Chinese architecture. Amazingly, the brick carvings in Bozhou Gorgeous Drama Stage remain intact after being exposed to the wind and rain for over 300 years without any protective measures. Without any doubt it has some secrets about the unique quality and techniques of carving materials. Brick carving has two advantages comparing to wood carving. It is easily preserved because its material is water and fire proof. It turns out that the blending of hairs and cotton fibers when making bricks will help them to be anti-erosive. It is the unique technological process that leads to the survival of the brick carvings with wonderful workmanship excelling nature till today.

Stylistically, the brick carving art is the combination of the fineness of Hui school and the vigor of Jin school. Terrazzo carving's blending of the two schools' artistic characters produces a mixed beauty of strength and fineness. Absorbing the artistic characteristics of Jin-style and Hui-style simultaneously, the polished brick carvings give a sense of stunning beauty mixing with hardness and softness.

It is a pity that the lost of skills of gray terrazzo brick carvings is confirmed by the experts of cultural relics management from Anhui Tourist Bureau. Therefore the existing ones are extremely precious.

Ⅲ. The History of the Construction of Bozhou Gorgeous Dramatic Stage

Nowadays, human activities and the modernization driving are nibbling the ancient buildings and historic relics in China. With its past splendor gone, the Gorgeous Drama Stage in the north of Bozhou is still attractive. Its brick carving, woodcut, luxuriant stage, colorful carved beams and painted rafter, keep its former glory as it used to be. This drama stage, which has a long history of 300 years, has a close relationship to *shaoyao* (the Chinese peony), Huatuo and herbal cultivation.

1. The Background of "the Home of Chinese Herbal Medicine"

Bozhou has a long history in trading and developing the traditional Chinese medicine. Since the East-Han Dynasty, the local people began to plant, process and operate the traditional Chinese medicine, which makes Bozhou "the Home of *shaoyao*" at that time. In the Ming and Qing Dynasties, as the local people began to plant *mudan*, Bozhou soon became the distributing center of the traditional Chinese medicine.

In ancient times, people could not draw a clear line between *shaoyao* and *mudan*. They both belong to Paeoniaceae and have similar shape and fragrance. However, the former is herbage while the latter xylophyta. *Shaoyao* is filled with treasure; what's more, its large flowers, bright colors and various species make it ornamental. The medicine made by its root is effective to spasm and sore. It was called flower fairy or flower prime minister, and one of the "six famous Chinese flowers". We also call *shaoyao jiangli* or *licao*, a kind of grass implying departure between friends or lovers. When it blossoms, it is splendid for its big and colorful flowers. Its leaves also have ornamental value. The "green dragon" in the verse

"Glistening is a seeming red lantern and crawling is a green dragon in outward appearance" is one of the praises to its leaves.

Before the Qin Dynasty, there were literal records of *shaoyao* but no records of *mudan*. From 770B. C. to 221B. C., a verse in *The Book of Songs*, the first poetry collection in China, "Lovers give the presents of *shaoyao* flowers to each other as a token of love". From this, we can guess that in ancient China, *shaoyao* was taken as a love keepsake in order to show people's love promise or reluctance to part. In this way, *shaoyao* was also called "jianglicao (grass that hints going away)" and now the symbolic flower of Chinese Valentine's Day.

Shaoyao is one of the earliest flowers cultivated in China. According to *On Mudan*, written by Xue Fengxiang, "*Shaoyao* first appeared and developed during the time of Xia, Shang and Zhou Dynasties, famous for its grace and elegance. The modern men admire *mudan* but look down upon *shaoyao*. They don't know *mudan* was nobody at the very beginning and was also mistaken as *shaoyao* for they have similar shape. "According to another ancient book, *Gu Qin Shu*, in the first year of Dixiang (the fifth king of Xia Dynasty), Tiaogu (a place) paid tribute to wutong and *shaoyao*. The king asked Yi to plant Wutong in Yunhe and asked Wu Luobo to plant *shaoyao* in the backyard. We know from this that *shaoyao*'s history for cultivation is more than 4000 years. In Bozhou, its history can be traced back to Western Zhou. In the early Zhou Dynasty, Bozhou was a part of Jiao. Jiao here, is considered the birthplace of the Chinese god in charge of agriculture, Shennong. In the age of early Western Zhou, the descendants of Shennong began to build cenotaph for him, set up medical center and teach the local people to plant herbs to cure illnesses. According to *Bozhou Annals*, the local people began to plant traditional Chinese medicine including *shaoyao* from 200A. D., which started a long history of planting *shaoyao* and other herbs in Bozhou. During the Three Kingdoms Period, Cao Pi, son of Cao Cao, introduced *shaoyao* especially in his *Emperor's View*.

In 200 B. C., the Chu Dynasty was overturned by Qin. After that, Qin set up Qiao County in the Bozhou Region, which gave Bozhou another name "Qiao". During East Jin, Emperor Cheng renamed Qiao County "Xiaohuang County". Liu Kai, a poet of the Qing Dynasty, wrote the poem "*Shaoyao* is blossoming outside

Xiaohuang County, just like the sunshine which covers five or ten li. The flowers are around local people's houses, every family plants rape flowers and mulberry trees." When he passed Bozhou and saw that there was *shaoyao* everywhere. From this poem, we can see that it was quite common to plant *shaoyao* in Bozhou then. Bozhou became the home of *shaoyao* at that time.

Mudan first appeared in the East Han Dynasty. The earliest record was a medical prescription discovered in a tomb of the East Han Dynasty in Wu Wei, Gansu Province in 1972 about how to use *mudan* to cure blood stasis. The ancient people began to transplant wild *mudan* since the Northern and Southern Dynasties. According to *A Prosodic Prose of Peonies*, "The emperor's hometown was in Xihe where a unique kind of *mudan* grew outside his house. The emperor asked his servants to transplant it to the royal palace when he found that there wasn't anything like that in the palace." *Mudan* became ornamental in the Northern and Southern Dynasties. It entered the royal palace during the Sui and Tang Dynasties. It is said in *Sui Yang Di Hai Shan Ji—Flowers and Trees in Palace* that the emperor took up an area of 200 li to set up Xiyuan and then collect rare animals and precious plants to raise in it. People in Yizhou (today Yi County in Hebei Province) presented him 20 different kinds of *mudan* in order to show their tribute. Emperor Yang of Sui also called *mudan* Sui Flower. In the Northern Song Dynasty, planting scale of *mudan* was unprecedented in Luoyang and the planting technology was greatly improved. More rare varieties of *mudan* were found. At that time, *mudan* was called Luoyang Flower or Capital Flower. The scale of *mudan* planting reached its peak in the year of Kaiyuan in the Tang Dynasty. The planting center transferred from Chang'an to Luoyang. Liu Yuxi, a famous poet in the Tang Dynasty, described the scene in his poem *Enjoy Mudan*, "Peony's colors can represent the nation. In its blossoming season, it moves the capital." During the Ming Dynasty, from 1368 to 1644, the planting center transferred the second time from Luoyang to Bozhou, Anhui Province. In the Qing Dynasty, it was popular in Cao Zhou (Heze, Shandong Province). According to the history book, *mudan*'s popularity began in Sui dynasty, prevailed in the Tang Dynasty and peaked in the Song Dynasty.

Xue Fengxiang, also named Gongyi, was born in Bozhou, Anhui Province in the late Ming Dynasty. He was one of the most famous horticulturists in the Ming Dynasty. He described the flourishing history of Bozhou *mudan* in the Ming Dynasty in his book *History of Bozhou Mudan*, which was the first monograph about *mudan* research in our country. It collects the past achievements on growing *mudan*, summarizes the cultivating skills and experiences of the *mudan* growers in Bozhou. This book proved that the cultivating and researching center of *mudan* transferred to Bozhou in the Ming Dynasty. It explains *mudan*'s varieties, cultivating skills, local customs and anecdotes.

When mentioning the cultivating history of *mudan* in Bozhou, Xue Fengxiang said in his *History of Bozhou Mudan*, "it was Ouyang Xiu who was once the administrator of Bozhou who should be to blame. He wrote no words about *mudan* as if there were not such a thing at that time. During the period between Zhengde and Jiajing, two of my late grandfathers named Xiyuan and Dongjiao were crazy about this flower and they hunted for it all over the country. From this time on, *Mudan* was cultivated in Bozhou." His late grandfathers mentioned above, Xue Hui and Dong Jiao, established a garden named "unique happiness" which was changed to "continual happiness" later. In this garden, different kinds of *mudan* were planted, including all the famous and precious varieties at that time. From this time on, *mudan* began to be cultivated in Bozhou.

Under the influence of Xue's grandfathers, the gardening of *mudan* became very popular when Xue Fengxiang grew up. When the late spring came, *mudan* was blossoming in all the great gardens and temples, rich in color. The unprecedented popularity was recorded in the *History of Bozhou Mudan*: The local people advocated *mudan* wholeheartedly, and there were few people who never grew the flower. They put their most beautiful flowers in a big bamboo tube with water inside, and made them compete with other's flowers in town. There were 14 great gardens which were famous for *mudan* in Bozhou. They differed from each other, raised by professional gardeners. Enjoying *mudan* became a fashion in Bozhou where many people cultivated, appreciated and researched on it.

The flourishing cultivation of *shaoyao* and *mudan* attributed greatly to the prosperity of Bozhou as the trading center of the traditional Chinese medicine. Hua

Tuo, a great doctor of the East Han Dynasty, took advantage of the herbs in Bozhou, the home of Chinese medicines. Hua Tuo (145 ~ 208), also named Yuanhua, or Fu, is an outstanding medical scientist of the late East Han Dynasty. He is one of the "three doctors of wonder in the East Han Dynasty", the other two being Dong Feng and Zhang Zhongjing. What he studied was the science of traditional Chinese medical. He opened up gardens to plant herbs and set up hospitals. He was called the earliest ancestor of surgery. He once visited Anhui, Henan, Shandong and Jiangsu etc. His contribution to traditional Chinese medicine can be divided into the following three aspects: Firstly, he invented a general anaesthetic Mafeisan and was the first person in China to perform operations. Secondly, by summarizing the theories of traditional Chinese medicine, he invented a system of therapeutical exercises called the Five Animals Exercises, which imitates the action of animals such as tiger, deer, bear, monkey and bird. Last but not the least, he invented artificial cultivating methods for herbal medicine according to the change of seasons. His efforts improved the quality of herbs. Fertile land and favorable weather made Bozhou an ideal place for planting herb. Herbal medicine became a type of main farming plants for the local people. Thus, Bozhou became a distributing center of the Chinese medicine, and now it still is.

Nowadays, Bozhou is one of the four famous traditional Chinese medicine centers. *Baishao*, a kind of *shaoyao*, was cultivated in Bozhou back in Wei and Jin Dynasties, and became a well-known herb throughout the whole country. Its yearly yield stands first in China. *Shaoyao* is the city flower of Bozhou. China's former Chairman Jiang Zeming wrote an inscription for Bozhou—Hua Tuo's Hometown, traditional Chinese medicine center. Every year on Sept 9th, Bozhou traditional Chinese Medical Trade Fair is held. Businessmen from all over the world gathered here to do business, to learn from each other, and to sightsee.

2. The Story of Building Bozhou Gorgeous Dramatic Stage

As a part of the Grand Temple of Guan Yu (an original name, another name of it is Shan-Shaan Guild Hall) in Bozhou, Huaxilou pavilion was built around 1756. The appearance of the guildhall symbolizes the sprout of capitalism and the financial prosperity of the society. Initialized in the very beginning of the Ming Dynasty, boomed in the 15th century of the middle Ming, prevailed in the Qing Dynasty, the guildhall served as a place to get together.

According to *An Introduction to Chinese Culture*, "The spreading and mergence of the cultures of different regions from the Jin to Song Dynasty exerted a great influence upon the prosperity of culture and economy of that period, which brings frequent exchange between the Northern and the Southern parts of China then." With the development of commercialization and the birth of capitalization, trade men all over the country gathered in cities and transportation fairs. For the purpose of doing business, merchants from the same place or of the same business often hold parties to exchange commercial information in a guildhall, so these guilds developed fast. There had been two kinds of guildhalls: one related to political and cultural issues, which was called "Townsman Club House"; the other was for the commercial parteners to exchange business information, which was called "Merchant Partner Club House". A "Townsman Club House" provided accommodation for students who came to travel, study, or take tests while "Merchant Partner Club House" was to hold business get-togethers. Mostly, the functions of those halls are not clear-cut. The merchants who did the same business got together there, especially in those "Shan-Shaan Guild Hall", which were of both cultural and economic functions. As a result of convenient transportation as well as regional and natural advantages, commodity economy flourished.

In Shanxi and Shaanxi, the two famous commercial groups became strong in the Ming and Qing Dynasties, named as "Jin Merchant" and "Qin Merchant" respectively, and "west merchants" if put together. Separated by a narrow strip of water, the two provinces enjoyed great intimacy with each other and the relationship was described in a Chinese idiom "The amity between Qin and Jin", meaning alliance between two families through marriage. At that time, the Jin-Merchants worked together to compete against the Hui-Merchants (a commercial group from Anhui Province) with unity. The Jin-merchant group was famous for its strong cohesion. They value the solidarity of family, folks and society. Though competing with each other, the fellows in this group helped and cared about each other as well. Shanxi merchants traveled from their hometowns to the prosperous cities in south

China. They transported necessities such as medicine, salt and tea from south China to north China. They sometimes stay in cities with convenient waterways in Yunnan-Guizhou plateau and eastern plains of the central China. Donated by the fellow townsmen, a guildhall was built to stipulate rules to deal with the business disputes and guarantee equal competition. In order to find a place in foreign lands to shelter themselves, share the business information and offer sacrifice to Guan Yu (who was believed by Chinese to be a god of wealth), the Jin merchants constructed many guildhalls all over China, known as "Shan-Shaan Guild Hall", for example, Sheqi Guild Hall in Henan Province, Liaocheng Hall in Shandong, and Xiangyang Hall in Hubei. In other places, Shanxi Guild Hall, Western-merchant Guild Hall and Qin-Jin Guild Hall were built up. Therefore, guildhalls served as a place for making friendly contracts, maintaining the group's interests, sharing advanced information, collecting donations, purchasing tombs and getting a rest for merchants outside their hometowns. It was also a union of self-governing, self-independence and self-control. With a strong solidarity, Shanxi and Shaanxi merchants had dominated the business circle for 500 years in China.

In the next five centuries, started with salt industry, Shanxi and Shaanxi merchants became business leaders in many industries, such as cotton, piece goods, foodstuffs, oil, tea, medicine, coat and finance. They expanded their business to many metropolis and trading centers all over the country, and even to other countries such as Mongolia, Russia and Korea, etc. Their activities enriched China's ancient business culture and pushed China's business to a new height. Their have shown their talents in both commercial and financial fields. They voluntarily donated lots of money to support the construction of guild halls. It was recorded that Shan-Shaan Guild Hall in Nanyang, with the title of Number One Guild Hall in China, used about 350,000 kilograms of silver in the first phase of the project and about 43,894 kilograms of silver in the second phase. One can appreciate the exquisite carvings and other kinds of architectural decorations in the existing Shan-Shaan Guild Halls all over the country. The superb and delicate technique is beyond this world. All of these show the great economic power of the Shanxi-Shaanxi merchants.

Bozhou Gorgeous Dramatic Stage, also named Bozhou Shan-Shaan Guild Hall, is one of the many Shan-Shaan Guild Halls built in the Ming and the Qing Dynasties.

As an ancient capital of the Shang Dynasty, Bozhou has been a key commercial port since then. In the Ming and Qing Dynasties, Bozhou was already famous for its production and trade of the traditional Chinese medicine, being one of its biggest distributing centers. Drug merchants came to set up shops there from all over the country. Drugstores could be seen everywhere in the city. About 170 different kinds of herbs grew in Bozhou. According to the Pharmacopoeia, *Bo Ju*, *Bo Shao*, *Bo Sangpi* and *Bo Huafen* named after Bozhou were four of the best medicines recommended by the doctors, especially the White *Shaoyao*, one of the three most famous kinds of *shaoyao* in China.

Bozhou is located in the central plain of China. To its north there was Guo River, which connected the Yellow River and the Huai River and enabled its convenient water transportation. In as early as the Spring and Autumn and Warring States Period, it was the distributing center of Chu, Song and Lu. In the Tang Dynasty, it became one of the ten most flourishing cities of the country. The commerce came to a peak in Bozhou when the guildhalls and shops sprang up all over the city during the Ming and Qing Dynasties. As the commercial distributing center of the Su, Lu, Yu and Wan, it was also called "the New Nanjing". According to *Bozhou Annals*, in the north part of the city near the Guo River, about 30% or 40% of the inhabitants were indigene, while the others were all their customers. All kinds of goods from the whole country were collected and sold here. Each street sold a different kind of good. Such as the street selling white cloth, bamboos, copper, dried fish, oxen and donkeys, etc. There were also hundreds of well-known drugstores in the city, such as Zhifang, Liren and Huashi, etc.

Bozhou Annals described the flourishing scene in detail: "The city was populated by rich businessmen; large ships were seen everywhere. You could see lots of delicate fabricates on the streets, and the business kept going day and night. When friends met each other, they got together to have a drink, play musical instruments, sing songs and enjoy tea. Though the prices were almost a hundred times higher than in other places, people were accustomed to it." Different trade associations from various regions set up guildhalls and native banks in Bozhou, such as, Guildhalls of

Huizhou, Jiangning, Fujian, Zejiang, Shandong, Henan and Jinlin, etc. Among them, Bozhou Shanxi and Shaanxi Guild Hall built in the north part of the city was the most magnificent one with the original name the Grand Temple of Guan Yu. Now we call it Gorgeous Dramatic Stage, for its religious and business function has faded away in recent years.

The main building of the Grand Temple of Guan Yu, Bozhou Shan-Shaan Guild Hall, was first built in 1656 (the 13th year of Emperor Shunzhi of the Qing Dynasty). In 1676 (the 15th year of Emperor Kangxi of the Qing Dynasty), a stage for drama shows was built in it. In 1766 (the 31st year of Emperor Qianlong of the Qing Dynasty), colored drawings and carvings were added to the stage. According to *Bozhou Annals*, this theatrical stage could not be more baronial and gorgeous. The two donators, Wang Bi and Zhu Kongling, from Shanxi and Shaanxi Province, were doing business in Bozhou. After they had great success in their businesses, they managed to build the temple. When established, the splendid and baronial guildhall became a wonder. After that, it was repaired for many times and turned out to be what we see nowadays. The repairing activities in the Qing Dynasty that can be traced are listed below: The first time in 1663 (the 2nd year of Kangxi); the second in 1694 (the 23rd year of Kangxi); the third in 1713 (the 52nd year of Kangxi); the forth in 1754 (the 19th year of Qianlong); the fifth in 1766 (the 31st year of Qianlong); the sixth in 1776 (the 41st year of Qianlong). After that, it was repaired twice in the years of Emperor Daoguang and Guangxu. The last colored drawing was painted in 1892 (the 18th years of Emperor Guangxu). Such repairs lasted for about 260 years, supervised by three Emperors. All these activities were done under the fund contributed by Shanxi and Shaanxi merchants. It is 351 years old now, having witnessed the striving history of Shanxi and Shaanxi merchants in Bozhou.

Bozhou Gorgeous Dramatic Stage was built in the Ming and Qing Dynasties, so it fully displays the building style at that time: magnificent in layout, and delicate in decoration, with strict building principals stipulated by the hierarchical system. In terms of its structural decoration, such as its color or layout, it looks like a palace. However since it was built by civilians instead of the emperor, it was restricted in some way in terms of decoration. If you look at the Grand Temple of Guan Yu in front of it, you can see that the roof is covered with gray tiles, not yellow glazed ones. Though the building is splendid, colorful and fully decorated, you cannot find any picture of dragon of five feet in it. These are because yellow tiles and dragons were only allowed to be used in the imperial palaces.

In the Chinese history of five thousand years, the country was dominated by absolute monarchy. The land of the whole nation belonged to the emperor, and all the people on the land were the emperor's subjects. Under the control of government, the monarch ruled the country depending on two things: protocol and law. *The Book of Rites* recorded that protocol was the order of a country. Traditional Chinese structures were divided into different ranks under the feudal system. The position, layout, shape, size, structure, material and ornament of buildings all showed strong political and moral norm, which was even written in the law by some emperors. As early as in the Qin Dynasty, the law about the ranks of buildings had come into being. Urban buildings where the king and dukes lived should be built according to the strict restrictions set by the law. If anyone didn't obey the building principal, he would be "violating the etiquette". When Duanshu in Zheng State arbitrarily expanded the size of his residence, he was criticized by a senior official Jishen as violating the etiquette and law. During the Sui and Tang Dynasties, the feudal system was more perfected; the rulers made a clearer stipulation to the patterns of the civil residence. The Song Dynasty's regulation of structural ranks even evolved in the limitations on building materials. *Building Patterns* divided building materials into eight ranks, stipulating that only the palaces of the highest rank with 9 to 12 rooms could use the first grade materials. Misuse of the materials was not allowed.

The Ming and Qing Dynasties had stricter limitations on structural grades. According to *Minghuidian*, houses of officials of different ranks should have different numbers of rooms, and different colors for decoration. In the Qing Dynasty, structures were divided into three categories: the palace, the large-scaled buildings and the small-scaled buildings. Palaces, where the emperor and his family worked and lived, were magnificent and gorgeous, with yellow glazed tiles and delicate structure. All types of colorful paintings could be used for decoration. Structures for officials, rich businessmen and retired officials were called large-scaled buildings and also beautifully decorated, but yellow glazed tiles or dragon and phoenix paintings were forbidden. The buildings for commoners were called small-scaled buildings, which are practicality oriented, seldom decorated and without any special structure. In addition, rigorous limitations were made for different parts of the building. The regulations showed the hierarchy in feudal society. Non-royal structures were all restricted by the rules of structural ranks in the Qing Dynasty. The houses built by Shanxi and Shaanxi merchants were unexceptional despite of their wealth.

The layout of the Gorgeous Dramatic Building was based on the characteristics of traditional Chinese structure. The layout was symmetrical and in order. In light of the layout of the building, neither the Bozhou Gorgeous Dramatic Stage nor the Grand Temple of Guan Yu is the exact name of the group of buildings. The Dramatic Stage and the Grand Temple of Guan Yu are merely two main buildings of the Bozhou Shan-Shaan Guild Hall. The Grand Temple of Guan Yu was built first, which was supplemented later with the Dramatic Stage. Like the Shan-Shaan Guild Halls in other places of China, the Grand Temple of Guan Yu represents the style of magnificence of Shanxi architecture, blending with the delicate style of Huizhou. The Grand Temple of Guan Yu is the core of the whole group of buildings, facing south to the Dramatic Stage. This is typical for traditional temple buildings in China. The hall of the temple faces south and is opposite to the Dramatic Stage that faces north and is connected with the gateway on its back. The gateway has three arched gates, with a gilded plaque engraved with the words "Grand Temple of Guan Yu". The Bell Tower and the Drum Tower stand respectively on the left and the right. A brick-paved passage connects the Dramatic Stage and hall of the temple. Whenever a drama was performed, the building structure could ensure a good sound effect. Such a delicate layout designed in the days when science and technology were not developed makes us marvel the wisdom of the ancients.

Bozhou Gorgeous Dramatic Stage (Shan-Shaan Guild Hall) is an important evidence of the existence of Shanxi merchants. It is not only a magnificent piece of architecture, but also an invaluable cultural heritage. The layout of the buildings and the decorative art of the Grand Temple of Guan Yu are of great significance to the study of the social, economic, cultural development of the Ming and Qing Dynasties, as well as the traditional culture and business philosophy of Shanxi merchants. The brick carvings of the Dramatic Stage shows the superb skills of ancient craftsmen, demonstrates the tremendous wealth and life philosophy of Shanxi and Shaanxi merchants and expresses the traditional Confucian thoughts.

Chapter

2

The Cultural Aesthetic Connotation of the Brick Carvings of Bozhou Gorgeous Dramatic Stage

Architecture is a kind of reflection of history. It represents culture, embodies wisdom and conveys people's expectations. Chinese traditional architecture tells China's 5000-year history and Chinese people's pursuit of happiness and a better life. From caves that sheltered the ancient people to the exquisitely decorated houses in the Ming and Qing Dynasties, the heyday of architecture and architectural decoration art in China, Chinese architecture developed together with the growth of economy, society and culture. It mirrors the characters of different dynasties, distilling their heritage. Bozhou Gorgeous Dramatic Stage (with its other names the Grand Temple of Guan Yu, Shan-Shaan Guild Hall, Bozhou *Huaxilou*) consists of a guildhall and a temple, which had economic and cultural functions in the Qing Dynasty. It is not a religious building isolated from the secular life, but a building closely linked with people's social life and living environment. Therefore, its brick carving, a representative of the exquisite decorative art of Bozhou Gorgeous Dramatic stage, reflects the changes of the eras, and all kinds of weathers in economy and culture.

We will focus on the cultural and aesthetic implications of the brick carvings on the main gate of the Grand Temple of Guan Yu, with a three-layered wood-like brick carving memorial archway. The building has three archways in the front. The bell tower and drum tower on both sides each has a small archway for decoration. In the middle of the archway is a plaque read *Can Tian Di* (meaning worshiping Heaven and Earth) and below it another plaque with the words "the Grand Temple of Guan Yu".

Around the two plaques the brick carvings are laid out orderly and symmetrically. The brick carvings on the Bell Tower and the Drum Tower correspond with those on the front gate, to achieve the aesthetic effect of coordination.

The brick carving of Bozhou Gorgeous Dramatic stage is the epitome of its kind in the Ming and Qing Dynasties. It also conveys profound culture which words cannot express. By using homophony, symbolism, association and metaphor, it conveys positive and promising ideas in Chinese culture, and reflects the ethics of traditional culture.

Ⅰ. The Perspective of "Everlasting" Philosophical Artistic Spirit

The Shanxi merchants, with their wealth accumulating and their political and social status rising, began to advocate Confucianism to call for a positive outlook on life. The cultural aesthetics of Shanxi merchants is reflected by the theme and design of the brick carvings in the Gorgeous Dramatic Stage.

There are three groups of brick carvings on the front gate of the Gorgeous Dramatic Stage. On the top of the main archway are *Flying Dragon Generates Rain*, Y*ingyang Banquet* and *A Flying Dragon and Swimming Fish*, with the first one at the highest position, representing the sovereignty of the emperor symbolized by dragon according to the hierarchy system in ancient China.

The second group of brick carvings is around the plaque of *Can Tian Di*, consisting of *The Three Stars of Fu*, *Lu and Shou Shining High*, *Bodhidharma cross the Yangtze River*, *Laozi's Alchemy*, *One Kui Is Enough*, *Kuixing Appointing Zhuangyuan* (the examinee who scored the highest in the imperial examination) and *The God of Wenchang*. This group is the soul of cultural connotation among all the brick carvings in the Dramatic Stage, for they represent the unity of Confucianism, Buddhism, and Taoism in Chinese traditional culture.

The third group of brick carvings below is around the plaque read "the Grand Temple of Guan Yu", including *The Warfare between the Kingdoms Wu and Yue* above and *The Whole Family Picture of Guo Ziyi's Birthday Celebration* below, with *Visiting Zhou Yu's Father-in-law at Ganlu Temple* and *The Taste of Life with Different*

Experiences on both sides. *Kuixing Appointing Zhuangyuan* and *the God of Wenchang* are on the sides of *the Whole Family Picture of Guo Ziyi's Birthday Celebration*. Below is a rectangular brick carving *Crane and Pine Tree* of *Longevity*. The semicircular decoration of *Two Dragons and a Pearl* locates at the lowest part.

All of the brick carvings mentioned above highlight the loyalty and bravery of Guan Yu, detail all seasons in life in a culture influenced by Confucianism, Buddhism and Taoism. Some brick carvings extol filial piety, the time-honored traditional virtue in Chinese culture. Another group of carvings is about four celebrities in ancient China who respectively loved chrysanthemum, lotus, goose and plum blossoms. Other brick carvings are about auspicious clouds and animals, which express the aspiration of the Chinese for a prosperous future. Scenes in historic stories were also carved for both decoration and admonishment.

The brick carvings on the top of the front gate are *Flying Dragon's Generation of Rain*, *Yingyang Banquets* and *A Flying Dragon and Swimming Fish*. This set of carvings generalizes and expresses the aesthetics and philosophy of Chinese traditional culture. The dragon is the symbol of the Chinese people and embodies the Chinese spirit. Dragon can generate wind and rain, which are beneficial to all the living things on earth. *Flying Dragon's Generation of Rain* takes up the highest position on the front gate, carving a vibrant dragon and a scene full of vigor. *Yingyang Banquets* presents a party held for those who excelled in imperial examinations after the examination, which was called the *Yingyang Banquets*. *Ying* means eagle in Chinese, symbolizing the top examinee would fly high and have a bright future. The carving advocates an enterprising spirit, encouraging people to struggle for a better life. It was also a manifestation of the optimism of Shanxi merchants whose economic and social status were rising. Besides, the content of the brick carvings is an artistic reflection of the inner worlds of the designers and craftsmen. The brick carving *A Flying Dragon and Swimming Fish* was based on an acrobatic drama, presenting a scene of fish and dragon playing with each other in the bright sunshine. Dragon was the symbol of emperors, it also implies joy and auspiciousness. Fish has the same pronunciation with surplus in Chinese, symbolizing bumper harvests and prosperity. The carving implies lively mental spirit and booming social development. The whole set of carvings present a scene of good weather, peaceful country and happy people. It reflects an optimistic attitude towards life and a thriving spirit.

The reason for that the Chinese culture can extend for several thousands of years and distinguish itself among all the excellent cultures is the life attitude of optimism and enterprise advocated by Confucianism. Confucius said, "Man should work so hard as to forget meals and be so optimistic as to forget the woes." As a man, one should be self-motivated and take an optimistic approach towards life.

In the Spring and Autumn Period, Confucius put forward the philosophical proposition of "knowing life but not death". Christianity soothes the suffered people by offering "the happiness in heaven" and "the call of God", instead of encouraging them to pursue earthly happiness. It leads people into the pursuit of the happiness in the afterlife rather than in this life. While the ancient Chinese philosophy had a good tradition, which did not pay much attention to the life after death nor establish moral on the immortality of the soul.

The Analects of Confucius recorded that Zilu consulted Confucius on ghosts and gods. Confucius said, "If one is unable to serve other people, how can he serve ghosts?" Zi Lu said, "May I take the liberty to consult you about ghost?" Confucius said, "I haven't thoroughly understood human affairs, let alone the affairs in afterlife!" Confucius believed that the real matter was to know life rather than death. Men live in the present, so they should base their moral values and pursuit on real life. People should work hard and be optimistic enough to live each day with passion. A sentence in a poem, "I work so hard, and do not feel the coming of old age", advises people not to worry about things after death or expect a happier afterlife, but to work hard in this life.

Since the Warring States Period, the Chinese have been inspiring themselves by the philosophical proposition "as Heaven's movement is ever vigorous, so must a gentleman ceaselessly strive along". "Vigor" was thought to be the nature of life. *The Book of Changes* said, "The great virtue of Heaven and Earth is creating life", "Continuous creation of life is change". It means that the principal nature of Heaven and Earth is changing. Since celestial bodies move vigorously and endlessly, people should be energetic and enterprising too. This kind of positive attitude is implied in the brick carvings of Bozhou Huaxilou such as *Flying Dragon's Generation of Rain*, *Yingyang Banquets* and *A Flying Dragon and Swimming Fish*.

Ⅱ. The Cultural Aesthetics of "Can Tian Di" and "Harmony"

According to Mr. Feng Youlan's theory of "philosophical characteristics", the Chinese traditional culture is based on Confucianism and complemented by Buddhism and Taoism. This culture has long developed since ancient times. The brick carvings on the front gate of the Gorgeous Dramatic Stage, are social, political and cultural symbols. They reflect the harmony between Confucianism, Buddhism and Taoism, representing the highest level of Chinese art.

Three Chinese characters, *Can Tian Di*, which means worshiping Heaven and Earth, are engraved in the plaque. The brick carvings are laid out around the plaque, including *The Three Stars of Fu, Lu and Shou Shining High*, *One Kui Is Enough*, *Bodhidharma Cross the Yangtze River* and *Laozi's Alchemy*. This set of carvings can be seen as the essence of all brick carvings, for it shows the unique cultural aesthetics of Chinese people and the talents of the builders. With its profound implication and rich connotation, it also shows harmony, the highest state of Chinese culture's aesthetic pursuit in a nearly perfect way.

The arrangement of the brick carvings around the plaque *Can Tian Di* represents the coexistence and harmony of Confucianism, Buddhism, Taoism in the Chinese culture. The three Chinese characters in the middle represent Confucianism which took up the dominant place in the Chinese culture, while *Bodhidharma across the Yangtze River* and *Laozi's Alchemy* on the sides represent Buddhism and Taoism respectively. Among these three schools, Confucians pursue the high-level virtue of being a saint; Taoists pursue the leisure and joy of "integrating with Heaven and Earth"; Buddhists pursue the dreamlike afterlife. The schools, representing three kinds of thoughts, each with its strengths and each playing its own part, have long nourished the spirit of the Chinese. With Confucianism playing a major role, they enabled people to find peace of mind in the complex world, and realized harmony between men and nature. In the thousands of years' development of the Chinese culture, Confucianism, Buddhism and Taoism advocate their respective ideas, showing all kinds of wisdom. They are the foundation of the Chinese culture and the spiritual home of the Chinese people. Chinese art consciously pursues the harmony between Heaven and Earth. This kind of pursuit can be seen all over the Chinese artworks in the old times.

The Warring State Period witnessed unprecedented academic prosperity, with numerous schools of thoughts contending one another. The Han Dynasty put forward "ousting all other schools of thoughts and adhering only to Confucianism", which decided the dominant position of Confucianism in the Chinese culture. Confucian culture is an ethical culture, which put virtue in the highest place. Harmony is the key to achieve virtue in this ethical culture. It has to be achieved among Heaven, Earth and people, between people and between one's body and mind.

From the perspective of social values, Confucianism plays a positive role in maintaining social order, for it extols virtue and ethics, advocates kindness and harmony. Taoism advocates learning from the spirit of irregular changing of the nature. It denies all the constraints of external form with wisdom. Buddhism advocates searching for one's own heart, adhering to inherent intuition, elevating oneself and relieving all beings from torment. On the level of spiritual life, Confucians pursue the self-improvement of the spirit and aspire to be a moral saint. Taoists advocate the liberation of the spirit and unworldliness and yearn for the state of "communing with Heaven and Earth freely". Buddhists aspire to spiritual purification and long for the realm of "nirvana". Actually, Taoists and Confucians want to achieve the same purpose by different means, for they both emphasize the harmony between people and the boundless universe. "Heaven and Earth coexisting with men and other things in the universe are of the same nature," read *Equality of Things by Zhuangzi*. Buddhism advocates broadening the mind, pursuing the perfect mental state, achieving holiness, truth and goodness and eventually reaching the state of "harmony" with nature and with others. Whether it is the order of Confucianism, the fine taste of Taoism or the Zen of Buddhism, they all play an important role in cultivating oneself as well as governing the country.

The set of brick carvings *Can Tian Di* uses symbolism to show that Confucianism,

Buddhism, Taoism coexist under the same sky. Though preaching their respective doctrine, they complement one another. The three beliefs are represented respectively by *Can Tian Di*, *Bodhidharma across the Yangtze River* and *Laozi's Alchemy*, which are put together. *The Three Stars of Fu*, *Lu and Shou* symbolize high positions, blessings and longevity in the Chinese culture. Confucianism, Taoism and Buddhism all show a unified concept of Heaven, Earth and people.

Located symmetrically on both sides of the plaque read *Can Tian Di*, are the brick carvings illustrating the story of *One Kui Is Enough*. According to ancient myths and legends, Kui is a monster, "it comes into and out of the water, generating wind and rain with the light like sunshine and the sound like thunder. Its name is Kui." read *Shan Hai Ching*. Another saying is that Kui was Emperor Shun's Yuezheng (an official in charge of music in ancient times). The ancients thought that music was the essence of Heaven and Earth, and it could unify the nation. Therefore, only saints could compose music. Because of his talent in composing music, Kui was thought to be a saint, so "One Kui is enough". The moral of the story of Kui is to give full play to people's talents. Since it was in the Temple of Guan Yu, it suggests that due to his loyalty and bravery, Guan Yu can be juxtaposed with Heaven and Earth. So one Guan Yu is also enough. For other people, they should also give full play to their talent to glorify Heaven and Earth.

The plaque *Can Tian Di* means to worship Heaven and Earth, which implies a pursuit of the unity of heaven and mankind. The notion of worshiping Heaven and Earth stems from the primitive Confucianism. In ancient times people cherished a sacred worship for Heaven and Earth. The Chinese civilization was based on self-sufficient and small-scale peasant agricultural economy in ancient times, so people often prayed for a good weather to enable their farming products to grow. Therefore the natural elements such as Heaven, Earth, the sun, the wind, the fire, the lightening and so on were all worshiped. The worships were categorized into three types: worship for Heaven and Earth, worship for ghosts and gods, worship for ancestors, among which the worship of Heaven and Earth is the most sacred. Not only the common people, but also the emperors and noblemen worshiped Heaven and Earth. The Temple of Heaven and the Temple of Earth now in Beijing are the sacred altar where ancient emperors showed homage to Heaven and Earth, praying for good weather and harvest.

Xunzi, the Confucian thinker in the Warring States Period and Sima Qian, the historian in the West Han Dynasty wrote these sentences in their works, "Heaven and Earth are the origin of life", "If there is no Heaven or Earth, where should people live?" Heaven and Earth are the natural environment people rely on and the foundation of all living things. If there is no heaven and no earth, how do people survive? So worshiping Heaven and Earth became one of the three basic elements for ritual together with enshrining the memories of the ancestors and revering the emperor and saints. The ancients believed that if they were to be blessed with an auspicious weather and a good harvest, they had to pay homage to the gods, Heaven and Earth. Out of worship and respect for nature, ancients believed that we should comply with the laws of nature, so as to be in harmony with Heaven and Earth to achieve the ideal state of "unity between nature and men". Confucius expounded the relationship between man and nature, "Only sincerity can make the best of man and nature, which helps assist the germination of Heaven and Earth. Only in this way can man be juxtaposed with Heaven and Earth." Confucius believed that only extremely sincere people can give full play to his own nature and the nature of all. Only in this way can man help Heaven and Earth foster life; only in this way can man be as great as Heaven and Earth.

The brick carvings on the front gate of Huaxilou has the three characters *Can Tian Di* (which means worshiping Heaven and Earth) in the center, reflecting the Chinese people's homage to Heaven and Earth, which highlights their wish to live in harmony with nature as well as their spirit of life. *Can Tian Di* can be regarded as the combination of "nature and man", implying that man can be juxtaposed with Heaven and Earth. *Can Tian Di* means that men should live in harmony with Heaven and Earth, and be inspired by Heaven and Earth to complete their life ideal. Guan Yu is the representative of such men. He is not only the epitome of morality, but also thought to be a guardian of the world with his loyalty and bravery, so he was comparable to Heaven and Earth. There is no doubt that the philosophy of worshiping Heaven and Earth provides spiritual support, confidence and strength for the Shanxi merchants in their struggle in business and life.

As mentioned above, the Chinese culture is usually thought to be shaped by Confucianism, Buddhism and Taoism. Thus it is also called "the culture of Confucianism, Buddhism and Taoism". Confucianism refers to the Confucian culture which is based mainly on the doctrines preached by Confucius and Mencius; Buddhism refers to the Buddhist culture; Taoism refers to Taoist culture or Taoists, the system of thoughts established by Laozi and Zhuangzi. It originated from Huanglaodao, a religion commonly believed in Chinese folk society in the East Han Dynasty, which had the Yellow Emperor and Laozi as its hierarch. It absorbed the idea of the Taoist theory about how to preserve good health and achieve longevity. The story of *Laozi's Alchemy* originated from Taoism, which is about how he refined the elixir of life to become a deity. We can see the image of "Lord Laozi", also called *Taishang Laojun* in Chinese, in all sorts of folktales, which was originated from the famous thinker Laozi. He was a deified figure in history, who was one of the most famous thinkers in the pre-Qin times and the founder of Taoist. Therefore, Taoists and Taoism both came from the same person, Laozi. But, after all, the doctrine and content of them are totally different.

Taoists believe that Tao is the origin of the universe and at the same time it is a miraculous thing with its own soul. The purpose of Taoists is to become deities, which lead to the immortality of the body and the soul. Both the Yellow Emperor and Laozi believed that only through inaction and stoicism can people realize Tao. Yet the Tao preached by Laozi and Zhuangzi, the founders of Taoism, is a concept of ultimate existence that cannot be defined. It is the source of all the endless and unlimited things. What Laozi and Zhuangzi advocated was the ideal state of "communing freely with Heaven and Earth". They advocated the philosophy of "letting nature take its own course", promoted the pursuit of spiritual freedom and liberation.

The artistic expression is often abstract and symbolic, especially in architectural decoration and painting, etc. When the images of saints of Confucianism, Buddhism and Taoism appear together in one building, one hall or even one painting, the boundary between the three religions becomes blurry.

Ⅲ. The "Loyalty Culture" of the Grand Temple of Guan Yu

Most of the guildhalls built by Shanxi and Shaanxi merchants during the Ming and Qing Dynasties were also temples for worshiping Guan Yu. They served as both a place of worship and a guildhall. The hall where Guan Yu was enshrined is called the great worshiping hall, or the main hall of the temple. The plaque in the middle of the front gate of Huaxilou (Shanxi and Shaanxi Guildhall) in Bozhou is inscribed with four Chinese characters "大关帝庙" (the Grand Temple of Guan Yu), showing that its original function was to worship Guan Yu. The dramatic stage was added to it later. "Plaque is the eye of the ancient buildings." The plaque integrates calligraphy into architectural decoration, which not only points out the name and function of the building but also can be decorative. The word "Grand" implies the magnificence of the building as well as the admiration for Guan Yu, the person who embodied loyalty in China. There is also a statue of the horse named "Red Hare" in the Grand Temple of Guan Yu, besides the statue of Guan Yu. There are other plaques, couplets, wall carvings and pictures in the building extolling the loyalty and integrity of Guan Yu. Why an ancient warrior deserves so much worship of the merchants? What is the connection between the spirit of Guan Yu and the business?

The Chinese civilization is one of polytheism. Three worships are integrated into the life of the Chinese, namely the worship of Heaven and Earth, the worship of ancestors and the worship of emperors and saints. Worship of Heaven and earth stemmed from the need of enabling environment; the worship of ancestors was human nature; the worship of the emperors and saints arose from the heartfelt admiration and spiritual needs. The "loyalty, integrity, kindness and bravery" of Guan Yu is one of the symbols of Chinese traditional culture in which two saints can be the model for the Chinese for thousands of years: Confucius and Guan Yu. The two respectively excelled in intelligence and martial arts, embodying the two mechanisms of Chinese governance. They are the spiritual pillars of the feudal society. Confucius was a famous educationist and the representative of Confucian culture. He belongs to the category of the sage, with the highest title of "the Accomplished and Sacred Master".

While Guan Yu was a warrior famous for his loyalty and patriotism and is worshiped as a guardian in folks. His highest title is "the Master of Great Loyalty and Extraordinary Bravery". These two are worshiped in temples built for them all over China. Guan Yu was extolled by Confucianism, Buddhism and Taoism for his loyalty, patriotism and bravery. Confucianism worships him as "the Martial Saint"; Buddhism refers him as "Sangharama Bodhisattva"; Taoism reveres him as "the Martial God of Wealth". Confucius is a civil saint, so the temples where Confucius was worshiped were called civil temples; Guan Yu is a martial saint, therefore his temples were called martial temples. The influence of Guan Yu far exceeds that of Confucius in the commoners. The worship of Guan Yu has become a part of their lives. Temples of Guan Yu, big or small, are everywhere in China's urban and rural areas. The status of Guan Yu has become so prominent that nearly everyone awes his power throughout China. He was thought to be as immortal as Heaven and Earth.

Guan Yu (162 ~ 220), with another name Yunchang, was born during Emperor Xi's reign in the Eastern Han Dynasty. As a native of Xiezhou (the nowaday Yuncheng in Shanxi), he was a renowned general under Liu Bei, the Emperor of Shu during the period of the Three Kingdoms (Wei, Shu and Wu). He was superior in martial arts and firm in character. Chen Shou's *The Romance of Three Kingdoms* said he was "brave and bold" and could resist "an army of ten thousand people". He had a "firm and restrained character". The Wei Empire Cao Cao, was deeply impressed by Guan Yu's bravery and his loyalty to Liu Bei, and wanted to persuade him to surrender and join the Wei army. But Guan Yu, determined to be loyal to his master, resolutely rejected Cao's offer. Though he beheaded several of Cao's generals, Cao still treated him generously and presented him with many gifts, including the famous horse "red hare". After that, to return Cao's appreciation and grace, he let go Cao in violation of the military orders when Cao Cao failed and fled on Huarong Road. The spirit of loyalty and bravery of Guan Yu was admired by people since then. The worship of Guan Yu has a lot to do with the three Chinese religions, namely, Confucianism, Buddhism, and Taoism. Therefore, the worship of him became religious. "Confucianism acclaimed him as Duke Guan to glorify his bravery and loyalty, hoping that all people could be faithful to others like Duke

Guan, so that there will be no fraud in the world. Buddhism regarded him as Sangharama Bodhisattva, who spreads his love, faithfulness and righteousness to the world; Taoism extolled him as one of the "five gods of wealth", who can bring people a prosperous future. Through the promotion of the emperors, worship of the commoners, the influence of the three religions and the extolling of *the Romance of Three Kingdoms* in the past more than one thousand years, Duke Guan has been sublimated from a historical figure into an icon of "moral and martial god" in Chinese people's minds and has become an emotional bond of Chinese descendants home and abroad.

In Shanxi, Guan Yu's hometown, people's worship of Guan Yu had reached the supreme status in the Ming and Qing Dynasties. So when people of Shanxi went out of their hometown to make a living, they prayed for the protection of the mighty divine, their "fellow townsman" Guan Yu. As we know, China's culture is determined by its agricultural economy. Since People were living a life of "working in the field early in the morning and coming back home at sunset, tiling the field and harvesting crops", a culture that favored farming over trading was formed. Before the Ming and Qing Dynasties, merchants were looked down upon, because the governments pursued the policy of "promoting agriculture and restraining commerce" and treated craftsmen as slaves and merchants as pariah. However, commodity economy developed in the Ming and Qing Dynasties and domestic and foreign trade prospered, resulting in a merchant dominated society, with money becoming a key factor in society. Therefore doing business became the best means of livelihood, even better than being an official.

Early entrepreneurs of Shanxi migrated to places far away from home, crossing deserts and rivers. The sinister business environment and the ups and downs in business rendered them helpless, so they could only pray to god in front of the unknown fortune. Therefore the loyal and valorous Duke Guan was chosen to be their patron saint who was thought to ensure them a thriving business. Guan Yu's righteousness had a clear cut difference from vulgarism and snobbery, with which people could live with each other without the conflicts of interest. This was an ideal situation favored by small and medium-sized producers in the feudal economy, an epitome of the lower people's interpersonal relationship. Guan Yu's bravery was thought to be able to rescue the weak and punish the evil. Guan's virtues were regarded as the principles in doing business, and practiced by the Shanxi merchants. Then Shaanxi businessmen joined them, forming the group called Western merchants. The merge of the two major business groups enabled Shanxi and Shaanxi merchants to run a thriving business in every commercial center and port all over the country. They donated money to built Shanxi and Shaanxi Guild Halls, where Guan Yu was enshrined. This greatly promoted the loyalty culture. The merchants of Shanxi and Shaanxi built the Grand Temple of Guan Yu to worship the saint for his "loyalty", "integrity" and "fairness". His "loyalty and good faith" had become the moral standards of the merchants, helping them to cultivate their own minds, constrain others and standardize the industry. At the same time the Confucian concept "if you are good at study, you will become an official someday" changed quietly, as some Confucian scholars started doing business. The core of Confucian culture is "benevolence", advocating "virtue". The spirit of Shanxi and Shaanxi merchants was in line with Confucius's doctrine. The influence of Confucianism can be found everywhere in the business culture of Shanxi and Shaanxi merchants. The "righteousness" embodied in Guan Yu's spirit was the virtue of traditional Chinese culture and was most advocated by Shanxi and Shaanxi merchants. The significance of building the Grand Temple of Guan Yu is to remind later generations that they should bear "righteousness" in mind every minute in their lives.

The couplets in the Grand Temple of Guan Yu, which read "the best model of righteousness the world has ever seen" and "adherence to the principle of righteousness and loyalty to the kingdom", are characterized by "righteousness". Businessmen, when doing business, paid special attention to amiability, as the saying goes "Amiability leads to wealth". There was also the saying "Failing to strike a bargain won't hurt the amiability". The merchants of Shanxi and Shaanxi advocated honesty and trustworthiness in doing business, making money in a righteous way and avoiding ill-gotten gains. They also preached never to forsake good for the sake of gold and never cheat in business, which embodied the spirit of maintaining harmony in fair competition. Righteousness, which referred to the fraternity between brothers and chivalrous feelings between friends, became the code of ethics among the merchants of Shanxi and Shaanxi in business and life. Ordinary people could not rival these merchants in righteousness and generosity. It was said that among Shanxi and Shaanxi merchants, if one store owed another a large sum of money but was unable to repay it, the creditor, for the sake of the debtor's self-esteem, let him only repay an axe or a basket and thus canceled the debt with a chuckle. This reflected the righteousness and generosity of the businessmen of Shanxi and Shaanxi. It also showed that people should have integrity even in difficult situation. Compared with Shylock, the selfish and greedy businessman in Shakespeare's *The Merchant of Venice*, the merchants of Shanxi and Shaanxi are noble and praiseworthy. It changed people's impression that businessmen were all dishonest.

Shanxi and Shaanxi merchants advocated Guan Yu's "loyalty and righteousness" also out of the need in doing business. China had the idea of "ruling the country with virtue" since ancient times. Contract and credit were very common in business, therefore businessmen had been keenly aware that a good faith was key to the success in business. Shanxi and Shaanxi merchants adhered to the integrity-based way of doing business, so as to earn the trust of the public and the society. The loyalty and integrity represented by Guan Yu had become the highest ethical standards for them to regulate the industry and protect business prosperity. These ethical standards enabled the thriving business of Shanxi and Shaanxi merchants to last for more than 500 years. Nowadays, "honesty" and "righteousness" are still the moral standard to adhere to if one wants to succeed. There is a remarkable similarity between Guan's loyalty and the honesty advocated by Confucians.

The Chinese have not only the virtue of honesty but also the virtue of "repaying", which means paying back the kindness they have received. Repaying is not only the traditional virtue of the Chinese but also an important principle and mechanism of moral life.

The brick carvings of Huaxilou, using a variety of methods, also promote "righteousness" and "repaying a debt of gratitude". Above the plaque read "the Grand Temple of Guan Yu" is "the *War Between the kingdoms Wu and Yue in the Warring States Period*". On the right side of the plaque is *Visiting Zhou Yu's Father-in-law at Ganlu Temple*. On the left hand side at the top is *The Oriole Repaying the*

Benefactor with Jade Bracelets. These stories and legends have a cautionary and enlightening function, admonishing people to be grateful to parents for their nurturing, to seniors for their guidance and to friends for their help, etc. Being ungrateful is immoral. Being grateful is an important part of Chinese moral conscience and a striking character of the Chinese morality. The brick carving "*The Warfare between the Kingdoms Wu and Yue*" located above the plaque read "the Grand Temple of Guan Yu" is to remind people that the market is like battlefield, where winning or losing is commonplace. One must learn to adopt a low profile and withstand setbacks in order to win the final victory. At the same time it tells people that merchants should learn from Fan Li, who did not covet undeserved reputation or money.

Ⅳ. The Traditional Chinese Culture Carved on the Brick Carvings

1. The Everlasting Chinese "Filial Piety Culture" from Confucianism

In the sequence of brick carvings on the main gate of Huaxilou, beneath the plaque "the Grand Temple of Guan Yu", there carved a classical brick work, *The Whole Family Picture of Guo Ziyi's Birthday Celebration*. It is the representative and essence of all the carvings on the archway for it has a splendid scene and the most characters among all the carvings. It conveys the information of filial piety culture which tops the ten traditional virtues and embodies the pursuit of the highest realm of Chinese Confucian culture "Harmony". This artwork has 42 characters carved with different poses, smiling or bowing. Guo Ziyi sat in the middle, with his long beard on his chest and a benevolent smile on his face. There is a Chinese character "寿" (shou) behind him. Many officials came to congratulate him on his birthday. On both sides of the carving stood two pavillons; the chariots and horses were running, showing a rich and peaceful scene. This blessed scene shows that the Chinese think high of the virtue of "filial duties". According to the data, the brick work *the Whole Family Picture of Guo Ziyi's Birthday Celebration* appeared in many building decorations all over the country. It is said that the copy of *the Whole Family Picture of Guo Ziyi's Birthday Celebration* in the Museum of Anhui Province is the best among all the brick carvings. It shows the Chinese tradition of "respecting the old and loving the young" is widely praised in China.

The Chinese have always regarded "filial piety" as the foremost of all virtues. Ancient Chinese thinkers were in favor of righteousness and universal love. It was proposed by Confucius that a gentleman should consider righteousness the most important virtue. A gentleman always thought that morality was supreme. Morality in ancient China included five items, of which benevolence was the top one. "Benevolence" originated from "sympathy", namely the compassion, which was based on the affection to one's family. The core of Confucius' benevolence, *ren* in Chinese, means "love of people", which refers to universal love. To begin with, one should love his parents. Therefore, the fundamental of benevolence was filial piety. According to *The Analects of Confucius* (*Lunyu* in Chinese), filial piety means, first of all, one should thank his parents for giving him life. He should feel happy for the health and longevity of their parents and concern about their aging. Filial piety is also about rewarding one's parents. The Chinese idioms "Crow's Feedback" and "Lamb Kneeling to Milk" told people to be grateful to their parents. Crow's Feedback goes that if a young crow grows up in the nurture of its mother, when its mother becomes old and blind, unable to look after itself, the young crow will look for food and feed it into its mother's mouth. This will last till the old crow dies. Li Shizhen's *Compendium of Materia Medica—Poultry Department* records, "For a crow, when it is born, the mother bird will feed it for 60 days. When it grows up, it will feed its mother for 60 days." In recent years, foreign zoologists studied the life of crows and they found out that crow has such a feedback behavior, while other birds don't. Ancient Chinese might find such behavior with their observation of the birds.

The "five conventions" in Chinese traditional culture generally refer to "benevolence, righteousness, propriety, wisdom, fidelity" (*ren*, *yi*, *li*, *zhi* and *xin*), which are regarded as the highest moral standards for ancient Chinese. Deng Zhongyue (1674 ~ 1748), an official in the Qing Dynasty, analyzed Chinese culture of five virtues in his essay brilliantly, "Pigeons chirp and crows feedback, which show benevolence. Deer sing to their group when finding grass and bees call on others when they see flowers, which show righteousness. Lambs kneel to milk and horses don't look down upon their mothers, which show courtesy. Ants dig holes to avoid water and spiders knit net for food, which show wisdom. Cocks don't crow until it is dawn, and swallows don't come back until it is spring, which show reputation. Even animals show these five virtues, let alone people, the soul of the universe!" Xunzi said, "Water and fire have spirit but no life; grass and trees have life but no wisdom; beasts and birds have wisdom but no righteousness; people have spirit, life, wisdom, and righteousness, so people are the most valuable from the ancient times." Since animals can do so many touching things, why can't people? People have their own life span, while by supporting their parents, being kind to their brothers, they can extend their virtues from generation to generation. From the point of religious beliefs and social values, Chinese culture is unlike the western culture which is attached to the theology or religions. The Chinese are polytheists, who do not actually believe in any of the gods. The Chinese are bound by moral. They think filial piety is the vital one. People's "feedback complex" is the most important force to maintain harmony in families and society. Disloyalty and unfilial are condemned for they will do damage to social harmony.

The similar stories about filial piety in the brick carving series of Huaxilou are *An Ape Steals Pantao for His Mother* and *Zeng Li Divorces His Wife* on both sides of the Clock Tower. *An Ape Steals Pantao for His Mother* is a legend. It was about a filial white ape. One day his mother wanted to eat *pantao*, a kind of precious peach, after recovered from illness. So the white ape went to Sun Bin's house to steal the peaches, but was caught by Sun Bin. The ape cried and told Sun its story. Sun Bin thought that it was a filial son, and presented it with some peaches and released it. The story illustrates filial piety is praised and practiced by the society generally. The carving works of *Zeng Li Divorces His Wife* is based on the story of Zengzi, one of the four major disciples of Confucius. Zengzi was known for his filial piety. He abandoned his wife just because she didn't steam the pear thoroughly for her mother-in-law. Zeng's behavior would seem unreasonable if the story happened in nowaday society, while it reflects that filial piety was a key part of morality.

Another carving, *Yanshan Educating His Sons* is about the father of the Dou family who made strict home discipline and educated his children in a proper way.

The name of the father was Dou Yujun. Being a native of the state of Yan by the Yan Mountain, he was also called Dou Yanshan. His five sons grew up in a good environment, in which "the father is benevolent, and the son filial; the elder brothers are friendly, and the younger ones respectful". As a result, all of the five sons had good achievements in the imperial exams. The story is a perfect model of good family education in Chinese culture. As the ancient saying goes, the sons will be filial only when the parents are strict to them. It emphasized the importance of the role of parents in the family. It shows how important family is for both ancient and nowaday Chinese. While the story *Wang Zhi's Rotten Axe Handle* reminds people especially the youth the value of time. It admonishes people to cherish time and life. These two stories both show great expectation to the young generations, encouraging them to make good use of their time and thrive to make a difference.

In addition, the thought of Confucian "norms and orders" is also reflected in the architecture and living arrangement in ancient China. Chinese traditional architectural culture was fastidious about the arrangement and the structural layout was often symmetrical. For example, to the east of the main entrance of Bozhou Gorgeous Dramatic Stage, there lies the Bell Tower, and to its west the Drum Tower. At the same time, the residential arrangements reflect the "filial piety culture". In a house, the older or the seniors live in the east room, the descendants or the young people live in the west room for the reason that Chinese people hold the east as the direction for the senior and the higher, while the west the direction for the junior and the lower. This becomes a set etiquette and a part of the filial piety culture. This kind of custom can also be seen in the arrangements of brick carvings. For example, *An Ape Steals Pantao for His Mother* and *Zheng Li Divorces His Wife* on the east Bell Tower correspond to *Yanshan Educating His Sons* and *Wang Zhi's Rotten Axe Handle* on the west side. The east side is the respected side, therefore, the brick carvings showing "filial piety" were placed on the east side, while the brick carvings about educating children were placed on the west.

2. The Business Culture of "Respecting Confucianism and Knowledge"

Shanxi merchants and Huizhou (Anhui) merchants were both regarded as Confucian merchants in history, for their ways of doing business reflected Confucian ideology and culture. Shanxi merchants were typical Confucian followers. They regulated their own behavior, and standardized the way of doing business by Confucian principles. Influenced by Confucian culture, in Bozhou Gorgeous Dramatic Stage, the works that embody the Confucian thoughts "righteousness" and "benevolence" can be seen everywhere. The construction of the Grand Temple of Guan Yu by Shanxi merchants showed their respect to Confucianism. In their business concept, "honesty" and "righteousness" are essential to success. The carving works *Three Visits to the Thatched Cottage* tells us that honesty and perseverance are needed if you want to invite a talent to work for you. *Madam White Snake* reflects the importance of righteousness from a different point of view. In the opinion of the Shanxi merchants, "justice" is not only a fundamental life standard, but also the essence of business. According to them, "Benefits come from the justice, and you shall get paid after you give." This showed that Confucianism was the essence of their thoughts. In the Qing Dynasty, Qiao Zhiyong, an outstanding merchant in Qi County, concluded the way of doing business: keeping credit, upholding justice and then making profits. His shop "Fusheng Oil Mill" became a representative for credibility in business. There was once his men were transporting a large sum of linseed oil from Baotou. Some of these men adulterated the oil for profit. When Qiao Zhiyong learned that, he immediately ordered to post notices to tell people about this incident overnight. Anyone who recently bought this kind of oil from his store could get a full refund. Although his shop suffered short-term losses, it gained reputation for Qiao's honesty. In Bozhou, Shanxi Merchants inherited and preached this good quality so that they could take advantage of their good reputation to attract more customers. Their story tells people to follow the principle of justice and mercy and the concept of righteousness.

Though Shanxi merchants' major task was to gain profit rather than educate the people, they strongly advocated the Confucian culture, which can be seen through the brick carving works in Huaxilou. Among all the 52 pieces of works, there are 20 pieces of works showing the Confucian culture, such as *Kuixing Appointing Zhuangyuan* (the Number One Scholar), *The God of Wenchang*, *Tianlu* (an imaginary animal) and *Kylin and Jade Book*. Kuixing was regarded as the god who dominated the destiny of essays in Chinese culture. In his right hand there is a large brush, which is used to appoint the person who passed the imperial examination. All the scholars worshiped him for his power to decide which the best essay was. The God of *Wenchang* (the God of literature and letters) was thought to be in charge of fortune and fame in Taoism. The image of *Kuixing* and *Wencheng* are often seen on big turtles and white tiger heads. In the pictures or carvings, *Kuixing* often points to the number one scholar using his writing bush, while the God of *Wenchang* holds a name list of the scholars in his hand. The implication was that if scholars studied hard, they would be blessed with a bright future. In recent years a kind of electronic dictionary occurred with the brand name of *Wenquxing*, which shows that modern students and scholars still respect the God of Wenchang.

The craftsmen carved these two main gods in charge of education in a prominent position. This shows Shanxi merchants' pursuit of high official positions and wealth. They changed people's prejudice against merchants that they were "mercenary and ignorant".

In Chinese feudal society for thousands of years, scholars were often considered noble and promising, for every year, a group of excellent scholars would be chosen by the imperial exam (*Keju*) to become officials. There were sayings like "all is inferior to studying", and "there is a gold house in books" to tell people the importance of studying. In China, children were educated to be hard-working from an early age. They were taught that "when reading, one should be deeply soaked in the book without any distraction". The scenes of hardworking students were described in some famous sayings: hang the head to the beam and stab the legs with needle to keep them from falling asleep; study hard for a decade in order to gain fame and wealth one day. So they "read hard for a decade just for that day when they could be selected as officials and granted a promising future.

Before the occurrence of imperial examination, officials were chosen by "hereditary system" (*Shiqing* and *Shilu* system), or "the policy of alien minister" (*Keqing* system). From the Sui Dynasty, the imperial examination system in China gradually changed. In the 7th year of the first emperor in the Sui Dynasty (587), Emperor Wen abolished *Jiupin Zhongzheng* system and began to select officials by tests. In his eighth year of reign, Emperor Wen set up two kinds of exams, which respectively served to select officials who had virtue and talent. On this basis, in the third year of the Sui Dynasty (607), Emperor Yang reformed the two exams into "Confucian classics" and "*Tien Si*". This is considered the beginning of the imperial examination, thus *Keju* system. From then on, Keju became an important method for ordinary people to improve their social status and earn a better life. The time when Shan-Shaan Guildhall (the Grand Temple of Guan Yu) was built in Bozhou in 1656, was just the most prosperous period of the imperial examination system in the Qing Dynasty. Although Shanxi merchants were engaged in doing business, they still revered the Confucian culture advocating study.

From the view of the psychology and value of the Shanxi merchants, as well as the business background at that time, Shanxi merchants' belief in knowledge might lead to more business opportunities. About the mercantile custom of Shanxi people, the governor of Shanxi Province, Liu Yuyi, wrote to Emperor Yong Zheng that, "Shanxi people live on doing business. Just as people know, they value fortune and wealth. Talented children often choose doing business for a life." Liu Dapeng in the Qing Dynasty said, "At that time, if a Shanxi family had a boy, the kid would learn to be a merchant instead of going to school. The parents thought that the more the boy read, the poorer he would be in the future, while being a merchant will bring him wealth. Therefore, the number of students to be examined was smaller than in other provinces in Shanxi Province."

The rank of occupations used to be, from the most favored to the least favored, "scholar, farmer, worker, merchant". However, since the smartest men became engaged in business in Shanxi Province, merchants began to move up the rank, and finally took up the place of the best occupation and changed the occupation rank to "merchant, scholar, farmer, and worker". The children from good families learned to read from childhood. They accepted good Confucian education, and then took up business. They turned into a group of Confucian elites active in Shanxi business domain. People used to think "if you are good at study, you should become an official after passing the imperial exam". While the Shanxi people advocated that "if you excel in study, you should be a merchant". From the modern point of view, paying attention to education has no contradiction with becoming a businessman. Education leads the Confucian spirit of integrity, righteousness and loyalty into business, so as to cultivate a merchant community with Confucian culture. This ensured that the Shanxi merchants were all elites and guaranteed that their businesses could prosper. The personnel management system and marketing strategies created by Shanxi merchants have been proved to be valuable, and worthy of learning by modern enterprises.

V. The Taoist Artistic Spirit Reflected by *The Drawings of the Four Loves*

On the front gate of the Grand Temple of Guan Yu are the carving works of *The Drawings of the Four Loves*, i. e. , *Wang Xizhi's Love of Geese*, *Zhou Dunyi's Love of Lotus*, *Duke Yin of Lu Watching Fishing* and *Tao Yuanming's Love of Chrysanthemum*, which are symmetrically placed on both sides of the gate. In the Chinese history, many scholars left us with not only precious literature or artworks, but also interesting anecdotes that reflect their noble character and fine taste of life. Wang Xizhi was a famous calligrapher of the Eastern Jin Dynasty; Zhou Dunyi, a philosopher of the Song Dynasty; Duke Yin, the Emperor of Lu State in the Western Zhou Dynasty and Tao Yuanming, a famous writer of the Eastern Jin Dynasty and a hermit. The four brick carvings are based on the anecdotes of the four celebrities. They are delicate in patterns, exquisite in skills, vivid in image, intriguing in plot and full of life interest.

Wang Xizhi (321 ~ 379, another record 303 ~ 361) was from Langxie (in nowaday Shandong Province). He was promoted to the second military chairman and secretary in charge of recording court meetings successively. He was a great calligrapher of the Eastern Jin Dynasty and one of the three greatest calligraphers in China. His cursive was considered the best in the Chinese history and he was acclaimed as the master of calligraphy. He had a passion for calligraphy all his life and loved raising and watching geese. After leaving the noisy capital city, Wang Xizhi came to Shaoxing, a pleasant river-town of beautiful scenes to the south of the Yangtze River. He often wandered in the garden, watching geese swimming on the water. Wang was inspired from the elegant poses of geese which helped him to refine his art of calligraphy. Wang loved feeding geese, about which many stories were told. A story goes that he exchanged a piece of his calligraphy for a flock of white geese. In exchange of the geese, he transcribed *Huangting Sutra* to a Taoist priest. Later people called Huangting Sutra "copybook in exchange for geese". The idiom "three-tenths of an inch into the timber" also came from a story of Wang. It is said that once he wrote a few words on a wood board and had a carpenter engrave the words in it. The carpenter, when carving, found that the ink left by Wang's handwriting permeated into the wood board for about three-tenths of an inch deep. Wang Xizhi's *Preface to Lanting Collection* is considered the masterpiece of running script. His calligraphy is as graceful as floating cloud and running water, extremely natural and unrestrained. In *Preface to Lanting Collection* there are 20 characters "之" (*zhi* in Chinese) in different writing styles, which looked like geese swimming on water, "some grooming its feathers, some looking around, all vivid and life-like." Emperor Wu of the Northern and Southern Dynasties commented that Wang's calligraphy was "as graceful as a flying dragon and a crouching tiger and therefore was valued for generations". Bao Shichen, a famous calligrapher in the Qing Dynasty said in *The Book of Jin—Biography of Wang Xizhi* that Wang's handwriting was so vigorous as if he had put all his strength into it.

Wang's love of geese is not merely admiring their appearances. It is out of the need of his calligraphic creation. By observing the geese's gestures, Wang Xizhi realized the essence of calligraphy and developed his unique style. At the same time, Wang forgot the troubles of life and enjoyed the peace of mind in the process of feeding and admiring geese. Calligraphy, as a unique art of China, reflects the artist's personality. "The style of calligraphy shows that of the man." Calligraphy is the reflection of the temperament of an artist. The charm of Chinese calligraphy comes from the aesthetic structure and pictographic characteristics of Chinese characters. Written on cocoon paper, *the Preface to the Lanting Pavilion Collection* presented the beauty of Chinese characters and the personality of the artist. It embodies the highest spiritual realm pursued by all the calligraphic artists.

The brick carving *Zhou Dunyi's Love of Lotus* relates the story of Zhou Dunyi, a master of neo-Confucianism in the South Song Dynasty, who loved lotus all his life. Zhou admired the nobility and integrity of lotus and adopted the qualities of lotus himself. He was an official of the Northern Song Dynasty, living a dignified and upright life. He once wrote the sentence "Lotus is not imbrued by the mud out of which it grows; nor is it coquettish though bathed in clean water" in his famous essay *The Love of Lotus*. Zhou compared lotus to a gentleman, claiming that "Lotus is the gentleman of flowers". A gentleman should have the noble character of lotus, but should not be aloof. He said he loved lotus in particular for its purity and righteousness. He praised lotus' unique and outstanding elegance and its chasteness though rooting in silt. Similarly, the Confucian culture advocates "self-supervision" for cultivating one's morality. A gentleman must introspect from time to time and never deceive himself. He must guard himself against evil will and indulgence. Even when he is alone, he must be honest to himself. Originally, lotus is regarded as a halidom of Buddhism, on which Buddha or Goddess of Mercy stands or sits, and a symbol of Dharma and purity. Zhou Dunyi, by explaining the image of lotus, succeeded in turning the lotus of the Buddhism into the lotus of Taoism and Confucianism. While in his view, self-supervision is far from enough in the complex and noisy society and a gentleman must have the quality of maintaining noble integrity like lotus remains pure despite growing out of tilt. The elegance of Zhou Dunyi's lotus is closer to the Taoist spirit, which pursues an elegant, quiet and ethereal life. Taoists admit and calmly face all that exist and advocate an open-minded and detached life attitude, through which Taoists free their hearts and liberate their spirits.

The brick carving *Tao Yuanming's Love of Chrysanthemum* is based on the seclusive life of the great poet of the Eastern Jin Dynasty. Tao Yuanming once acted as the chief of staff in Jiangzhou and staff officer in Zhenjun. But he was dissatisfied with the reality and had no interest in fame or wealth, so he resigned and lived in seclusion on a farm. He loved chrysanthemum throughout his life. After retirement he plowed everyday in the chrysanthemum garden. His famous poem, "Pick a chrysanthemum near the east fence, and leisurely I see the mountain in the south" reflects his life attitude after his return to the nature. Another poem of his, "Though the paths in the garden have nearly been wasted, the pine trees and the chrysanthemums are still there" is the reflection of the poet's life. Having abandoned high position and great wealth and avoided socializing with high officials, the poet achieved the peace of mind in his rural life. Tao Yuanming loved chrysanthemum mainly because it symbolizes his own personality. As Zhou Dunyi commented, "From my perspective, chrysanthemum is the hermit of flowers." In the bleak autumn, chrysanthemums are in full bloom, braving the frost. It symbolizes the poet's rebellious character. Tao not only abandoned the fame and wealth of an official, but also, unlike other hermits, began the life of a farmer. This is the most unique characteristic of a hermit, i.e., "communicating independently with the spirit of Heaven and Earth but not despising other things" from *Zhuangzi*. By interacting with nature, his personality reached the realm of "carefree leisure", which has always been the Taoists' pursuit.

The other three of the four figures in the *Drawings of the Four Loves* are men of letters, while only Duke Yin is an emperor. Why was *Duke Yin of Lu Watched Fishing* chosen as one of the four loving? We can get some hints from the history of the State of Lu. Ji Boqin, the first king of the State of Lu, was the eldest son of Duke Ji Dan, who made all the rules and etiquette of the Zhou Dynasty and later retired and returned to Lu. Thus Lu was granted the privilege of offering sacrifices to Duke Ji Dan with the etiquette of a monarch, and Lu had the closest relationship with the central government and enjoyed the highest status among all the kingdoms. At meetings, the status of kingdoms is decided by the spirit of patriarchal clan and Lu ranked the first among all kingdoms with the surname of Ji. The State of Lu was dubbed as "the state of ceremony" which retained the most complete rites of all the kingdoms, and its national history was the most complete. Confucius, the founder of Confucianism, was born in Lu. The scripture of the time-honored *Spring and Autumn Annals* was adapted by Confucius according to the national history of Lu. *The Spring and Autumn Annals* began from the first year of Duke Yin's reign (722 B. C.), which made Duke Yin known in history. Duke Yin's story was most representative.

Duke Yin of Lu, with the surname of Ji, was the thirteenth ruler of Lu in the Zhou Dynasty. Why was he called Duke Yin? It was a long story. Ruler of a state was usually addressed as "Duke" in ancient times, implying respect. And Yin was his posthumous name. In ancient China, an emperor's name was often addressed together with his posthumous name, which was given to him after his death, implying a judgment from his later generations. The posthumous name occurred in the Zhou Dynasty and was abolished in the Qin Dynasty, but in the Western Han Dynasty it prevailed again. Duke Ji Dan and Jiang Ziya made great contribution to the Zhou Dynasty and was given posthumous names after their deaths. This was the beginning of posthumous name. *Zhou Rites* said, "An official was given a posthumous name soon after his death." *The Book of Zhou—Regulations for Posthumous Title* said a posthumous name was the trace of a man's feats. Great deeds beget great titles. So a man's posthumous name was earned by the man himself. The regulations for posthumous title of the Zhou Dynasty stipulated that a posthumous name must match the personality of the deceased. The posthumous name of an emperor was decided by officials in charge of rites and was announced by the successor of the deceased emperor. The regulations for posthumous title defined some words of fixed meanings, which could be roughly divided into three categories: firstly, words with positive meanings, such as, *wen*, *wu*, *ming*, *rui*, *jing*, *kang*, *zhuang*, *xuan*, *yi*, *lie*, *zhao*, *mu*, etc.; secondly, words with derogatory meanings, such as *yang*, *li*, *ling*, etc. To take Emperor Li of Zhou as an example. The Emperor killed many innocent people wantonly, hence was given the posthumous name Emperor Li. Third, words denoting sympathy, such as *ai*, *chu*, *min*, *dao*, etc. For example, the Emperor Huai of Chu was called such a name for his kindness. Later Wang Guowei, the famous scholar of the Qing Dynasty drowned himself in June 1926. Puyi, the last emperor of China, granted him the posthumous name of "zhongque" (que means honesty), so his tombstone was engraved with "Mr. Wang Zhongque". This marked the end of the posthumous name system in China.

Here we will tell a story about Duke Yin's name. Yin's father, Emperor Hui of Lu had no son at first, but then his concubine Shengzi gave birth to a son named Zixi. When Zixi grew up, his father, the emperor, found a wife for him from Song State. However, his father found the Song girl beautiful and made her his own wife, who later had another son, Yun, Yin's step-brother. Emperor Hui made the girl queen and her son the crowned prince. When Emperor Hui died, Duke Yin acted as the regent because Yun was too young to accede. So Duke Yin administered the state but had Yun (later called Emperor Huan) as the emperor, which made him an almost invisible person. Thus his posthumous name was Duke Yin. The rituals of the Zhou Dynasty were complete and strict, being the essence of its governing. Due to his status as the regent, Duke Yin of Lu kept a low profile all his life, strictly observing the rituals in many occasions.

The story *Duke Yin of Lu Watching Fishing* is based on an episode named *Zang Xibo Admonished Duke Yin against watching fishing* in *Zuozhuan*, the first detailed narrative chronicle in ancient China. It is a detailed account of Duke Yin going to Tang (in nowaday Shandong Province) to watch fishing in the spring of his fifth year in power, which was later criticized by Zang Xibo. According to the customs at that time, fishing was a base trade. As the monarch of the state, he should be committed to something important, such as sacrifice, development of national defense and the governing of the country. A monarch should be restricted to rituals in words and deeds, hiding his individuality, giving up his freedom to become a role model in observing the rituals. As a monarch, Duke Yin's watching fishing was not in line with the rituals of Zhou. But Xibo failed in dissuading Duke Yin, so the latter did go to Tang to watch fishing. Xibo pretended to be ill and didn't go with him.

Why did Duke Yin of Lu not listen to advice and persist in going to watch fishing? Liu Zuqian, a man in the Song Dynasty unraveled the secret, "Every emperor enjoys the pleasure of touring and feasting and hates serious advice. So it is difficult for an emperor to heed advice if it is to take away his pleasure." Xibo's advice was just to take away Yingong's pleasure. Entertainment appeals to the heart of everyone, and kings are of no exception. "After a long time in the cave, a bird aspires to get back to nature." Duke Yin of Lu's indulgence in pleasure and pursuit of the spiritual freedom reflect his true personality. Although Duke Yin of Lu lived about 100 years earlier than the appearance of the theory of Laozi and Zhuangzi, he behaved in a way that was in line with the life interest of Taoism. What he ignored was the rituals from Confucian, which originated from a land near his hometown. To be honest, the ideas of Confucian and Taoist are not contradictory if we look at them from a nowaday view.

Ⅵ. The Cultural Aesthetic Message in "Auspicious Clouds and Animals" of the Brick Carvings

"The basic characteristic of Chinese art is determined by the characteristics of Chinese culture." According to ancient Chinese culture, all things are derived from *Qi*, a kind of spirit. When the spirit is gone, the thing dies, returning to void. The sun, the moon and the stars in the sky, mountains, rivers, animals and human beings on Earth all have spirit. Spirit is the essentials of the universe and the basis of all the things. So it is of course the basis of art works. Therefore, all the things depicted in the artworks need to be lively, full of *Qi*, which is the basic character of the Chinese art. Among all the things carved, the most lively ones are animals and plants. These living beings symbolize ancient Chinese's wish and pursuit.

In the brick carvings of the Gorgeous Dramatic Stage, many animals have special meanings. For example, on the main gate of the Grand Temple of Guan Yu, there are some carvings using homophones to mean something different, such as *Nine Lions Playing the Ball* (*Jiu Shi Tong Ju*, meaning nine generations live together) and *Five Lions Playing the Ball* (*Wu Shi Qi Chang*, meaning five generations live together); *The Whole World in Spring* (*Liu He Tong Chun*) and *Three Yangs Bring Auspices* (*San Yang Kai Tai*). There are many other images that imply the beautiful wishes of the Chinese people, such as *Two Dragons Playing with the Pearl*, *Dragon and Phoenix Indicating Good Fortune*, *A Rhino Watching the Moon* and so on on the main gate. Brick works on Bell Tower and Drum Tower, such as *Pine Trees and Cranes*, *Longevity as Pines on Zhongnan Mountain*, *Phoenix and Lotus*, *Mandarin Ducks Playing with Lotus* and so on use plants and flowers to indicate the happiness and joy in life. Let's take *Nine Lions Playing the Ball* as an example. It means nine generations live together in a big family. In Chinese, "lion" and "shi" have similar pronunciation. Therefore the carving uses the image of lion to represent "generation". Lion is also taken as an auspicious animal because it is fierce and brave, being one of the top predators. In Chinese artworks, little lions are often depicted lively and lovely, and the big ones mighty and strong. Lion has its other names in China, such as the Buddhist name *Suanni*, for it is regarded as the Dharma beast in Buddhism. Another name is *Yi Pin Chao*, which has something to do with politics. In Chinese culture, *Pin* was the official rank, from one to nine. The first *Pin* was the highest. Since the lion is the king of beasts, it was regarded as the first *Pin*. As mentioned above, the nine lions here symbolize nine generations. Therefore here is the meaning of the carving: nine generations are living together happily. Having a large family was the wish of all the ancient Chinese people. They felt proud when their children and grandchildren are around, for they thought many children and grandchildren would bring them happiness. Therefore, a big family was thought to be "blessed". The English phrase "extended family" can convey the Chinese people's desire best. In such families, the blood of the old generation could pass down. It was common for a family in old times to consist three or four generations. While five generations live together was rare, not to mention nine generations.

The Whole World in Spring (*Liu He Tong Chun*) and *Three Goats Bringing Auspices* (*San Yang Kai Tai*) also use homophone "six" and "deer", "he" and "crane", "yang" and "goats". A deer and a crane in the carving indicates the spring and the vitality in the world, when man can enjoy the warmth and energy. In Chinese calendar, the first month of lunar calendar is "yang"; "tai" means changes in the world, here refers to the coming of spring and the departure of winter. So, *Three Goats Bring Auspices* represents good wishes at the beginning of a new year. An old couplet *San Yang Kai Tai* (*Three Goats Bring Peace*) and *Wu Shi Qi Chang* (*Five Generation Flourishing Together*) expressed people's hope for a flourishing family and peace in the world in the coming spring. You may ask why Chinese people paid so much attention to the change of seasons. This was because China was an agricultural country, and Chinese people lived in an agricultural economy. At that time, science and technology were underdeveloped, natural disasters often occurred and ruined their agricultural production. Therefore, their survival depended on good weathers. It was said in *Zhouyi*, that "No harvesting, no wealth." They prayed for God to "open his eyes" and be kind to protect their harvest and safety year after year. As a saying goes, "Spring is the best season of a year", the Chinese attached great importance to the beginning of a year. Spring is also the start of farming. "Planting in the spring and harvest in the autumn." "A good spring planting indicates a good harvest in autumn." Therefore, to convey the wish for harmony between man and nature, *The Whole World in Spring* (*Liu He Tong Chun*), *Three Goats Bringing Auspices* (*San Yang Kai Tai*), and *Every thing updates* (*Wan Xiang Geng Xin*) are carved.

The art of brick carvings expresses its deep ideological content by image. It is different from other art forms such as literature or paintings because it needs to be simplified due to the limitation of materials and the area. The carving works, such as *Two Dragons Playing with the Pearl* (Er Long Xi Zhu) and *Dragon and Phoenix Indicating Good Fortune* (*Long Feng Cheng Xiang*), and other decorations on the front door of the Gorgeous Dramatic Stage, the Bell Tower and the Drum Tower, used the simplification and varieties of forms to express the meanings. Dragon and phoenix are both mythical animals symbolizing happiness in Chinese culture. They often show up together in architecture decorations. The book *Rituals and Forms* in ancient Zhou Dynasty stated, "Unicorn, phoenix, turtle and dragon are four animals of intelligence." Except the turtle, the other three are all mythical. They can symbolize the pursuit of peace and happiness, good fortune and longevity.

The image of phoenix originated from ostriches in the Stone Age. The legendary goes that ostrich could distinguish music and was good at dancing. Therefore, the phoenix was regarded as a female image of morality. From ancient times, the dragon was associated with the emperor. Since the beginning of the Han Dynasty, the first ruler Emperor Gao, claimed that he was "the son of the god, the dragon" in order to stabilize his position. From then on, the image of dragons exclusively referred to the emperors in feudal society. Besides that, dragon is a symbol of joy and strength. In order to achieve the match of *yin* and *yang*, people found phoenix as the match of dragon. Dragon stood for *yang* and masculinity, while phoenix represented *yin* and feminity. Therefore, we often see art images of dragon and phoenix in ancient architectural decoration.

Kylin, also named *qilin* with *qi* referred to the male and *lin* the female, looks like a unicorn. It was believed to have moose body, ox tail, wolf hoof, and a horn.

While it was recorded as "with horse feet, ox tail, yellow, round hoofs and a horn" in *Mao Shi Yi Shu*. It is said to be an imaginary animal originated from deer. It was also very likely, a totem of Zhou clan. The ancients thought the deer was "the beast of pure goodness". Deer was thought to be a symbol of purity and morality. So there was a saying "a moral person can see white deer". Kylin symbolizes an outstanding figure. People once called clever kids "Kylin boys" in the past.

The Chinese culture emphasizes "the harmony between man and nature". People's spirit and behavior should be consistent with nature. They strive for the balance between body and mind, and the balance between man and nature. Therefore, Chinese art often relate other things to human in nature to get "harmony". For example, in Chinese literature, plum, orchid, bamboo and chrysanthemum are acclaimed as "four gentlemen" in flowers because of their elegant characters. Pine, plum and bamboo are called "three friends in winter". These plants are often used to refer to men with dignity.

Since ancient times, there are uncountable poems and essays praising the character of plum blossom. Mao Zedong described the character of the plum blossom with an optimistic attitude, "Wind and rain send spring here, falling snow welcomes it. On the iced rock, only blooms the pretty plum blossom. Sweet and fair, she does not crave spring for herself alone. She only tells us that spring is here. When the mountain flowers are all in full bloom, she smiles in their midst". But in Lu You's poems, he depicted the plum blossom with desolation and loneliness, "Beside the broken bridge out of the stage, plums open lonely. Saddened by her solitude in the falling dusk, she is now assailed by wind and rain. She has no intention of bitter strife, letting other flowers be envious. When it falls in to Earth, it becomes the dust, with only fragrance remains". On the right-bottom of the Drum Tower, the brick carving *Plum Blossoms and Magpies* features a pair of magpies perching in a plum tree in full bloom. The flower looks so real as if you could smell its fragrance. One of the two birds is singing, and the other is pecking. Plum flowers coming out at the beginning of the year are called "primrose". Therefore, a pair of Chinese Spring Festival couplets read, "Spring is the beginning of the year, plum blossom is the queen of all flowers". The magpie has pleasant chirps, so it gets its nickname

"Bono". There is a folk saying "When you hear magpie chirp, it means good luck is approaching". Since plum blossom and eyebrow are homophones, a magpie on a plum branch means happiness will appear on the eyebrows.

Pine and cypress often show their distinctive characters during the cold season. In *the Analects of Confucius*, Confucius said, "The pine and cypress are evergreen during cold season." Zhuangzi said, "When cold comes and snow drops, I know the booming of the pine and cypress." The ancient Chinese regarded the pine and cypress as symbols of long life, for they were evergreen and resistant to cold and hardship.

Crane is a kind of long-lived bird. The Chinese word "crane age" is used to refer to the age of an elderly. So pines and cranes have identical meaning of longevity. Peony, Flower King in China, is loved by Chinese because of its beauty and magnificence. Peony symbolizes prosperity, splendor, wealth and happiness. Lotus, in the description of Zhou Dunyi, a great litterateur in the Jin Dynasty, is a "gentleman" in flowers coming out of the mud, but not imbrued. It also refers to a pure and beautiful lady. The carving works on Bell and Drum Tower, *Pine and Crane of Longevity*, *Longevity as Pine of Zhongnan Mountain*, *Phoenix and Peony*, *Mandarin Duck with Lotus*, etc., all imply the meaning of good luck, longevity, reunion and blessing. They all reveal the Chinese cultural characters of profound connotation.

第三部分

亳州花戏楼砖雕故事文化图解

正门牌坊
Carvings on the Main Gate

1. 鹰扬宴　位置:正门牌坊上层花板(西)

图中上刻一只展翅腾空的雄鹰,下刻一雁,因"雁"与"宴"谐音,故名"鹰扬宴"。

此图的涵意来自于中国古代科举考试制度。科举是中国封建社会选拔官吏的一种考试。它始于隋炀帝大业四年(607 年),一直沿袭到清光绪 31 年(1905 年),历时 1300 多年。科举制度是古代中国的一项重要制度,对中国社会和文化产生了巨大影响。现代社会公务员的选拔制度也是从科举制间接演变而来的。

科举考试分为三级:乡试,会试和殿试。乡试即省级考试,参加者为府、州、县考试中举的秀才。乡试后翌日,通常要为乡试中举者举行由朝廷主办的盛大庆祝宴会,以示恩典。自唐代以来,科举考试分设文武两科。因此形成了我国古代著名的科举四宴:鹿鸣宴、琼林宴为文科宴;鹰扬宴、会武宴为武科宴。

此幅《鹰扬宴》寓意着为武科状元举行的宴会。"鹰扬"形容威武,寓意考中武举者将会像威武的雄鹰一样前途无量。这幅作品目的是鼓励人们积极进取,努力奋斗,以求功名。

◉ **Yingyang Banquets**

This picture represents a flying eagle at the top and a wild goose at the bottom. The Chinese equivalents of wild goose and banquet are of the same pronunciation *yan*; thus the carving was named *Yingyang Banquet*.

The imperial examination serves as a tool of selecting feudal officials, dating back to the fourth year of Daye, the reign title of Emperor Yang of the Sui Dynasty (607). In the time of more than 1300 years, it continued until the 31st year of

Emperor Guangxu of the Qing Dynasty (1905). Official celebrating feasts are presented regularly for the exam passers of wisdom and valor whom the rulers invite to their sides, showing the emperor's grace. They are composed of banquets of Luming, Qionglin, Yingyang and Huiwu. Imperial examinations were divided into civil and military service examinations since the Tang Dynasty, and the feasts Luming and Qionglin were provided for civil service examination passers, while the feasts of Yingyang and Huiwu were for military service examination passers. This piece of artwork aims to encourage people to strive for higher social status.

2. 龙腾致雨 位置:正门牌坊上层花板(中)

龙是中国古代传说中的一种有灵性的动物,为麟虫类之长。据说龙能兴风致雨以利万物,故为四灵之一。四灵是苍龙(青龙)、白虎、朱雀和玄武。中国是一个古老的农业社会,万物的生长都要依靠风调雨顺,因此在古代中国,龙的形象被认为具有较高的地位。中国古代的统治者都自喻为"真龙天子"。龙是一种高贵神圣的艺术表现主题,多用于建筑、雕刻、服装、刺绣等艺术装饰中。

龙是中华民族古代图腾的一种,中国人以"龙的传人"自称。据传,古人为研究人类生育过程而将孕妇的子宫剖开,发现三至四个月大的胎儿长有尾巴,觉得不可思议。开始人们以为这是巧合,后证实母腹中胎儿都有尾巴。于是,人们便认为三四个月的胎儿是一种神灵,而后才在母胎中变成人。古人根据图腾崇拜,给这种神灵取名"龙"。于是中国人是"龙的传人"的说法得到普遍认可并流传下来。

此幅砖雕图刻一游龙于祥云之中,居于整个砖雕组图中央。"龙腾致雨"寓意着风调雨顺、国泰民安。

◉ Flying Dragon's Generation of Rain

According Chinese legend, dragon is a miraculous animal that can generate cloud and rain to benefit all the things on Earth and thus is acclaimed as the head of four animals with intelligence. The four animals are Qinglong (dragon), Baihu (tiger), Zhuque (phoenix) and Xuanwu (tortoise). Dragon was highly valued in agricultural society because the survival of everything on Earth depends on timely rain. So it enjoyed a high position in ancient China.

The ancients were curious about the reproduction of human beings so they cut open a pregnant woman's womb to see what the fetus looked like. They found out that fetus had a tail, which surprised them. Initially it was supposed to be a coincidence, but then it was proved to be universal after they had seen into more pregnant women's womb. They tended to believe that the 3-4 month fetus with a tail was a kind of supernatural being which would then grow into a human being. According to totemism, people named the fetus "dragon". The idea that human is dragon's offspring became widely accepted then and grew popular.

This sculpture is situated in the center of the carving. It embodies the good wishes of ancient Chinese for timely rain for the crops to grow so that the country could be prosperous in which people could live peacefully.

3. 鱼龙漫衍 位置:正门牌坊上层花板(东)

此幅砖雕名为《鱼龙漫衍》,上刻一巨龙腾云驾雾,下刻一比目鱼在水中嬉戏。

在古代,"鱼龙漫衍"或"鱼龙曼延"是一出杂耍戏的名字。戏中鱼龙为传说中的巨兽,"先戏于庭极毕,乃入殿前激水,化成比目鱼,跳跃漱水,作雾障日毕,化成黄龙八丈,出水敖戏于庭,炫耀日光"。"漫延(衍)"意为"巨大",而鱼龙、龙、比目鱼在中国都是瑞兽,人们相信它们能够带来好运。"比目鱼"在古代中文中的意思是成双成对游动的鱼。古代人们认为比目鱼的两只眼睛长在一边,所以游动的时候需要两条同类别的鱼一起游才能辨别方向,于是赋予了比目鱼成双成对的含义。后以比目鱼比喻形影不离的人或情侣,故比目鱼又被人们看作爱情的象征。古代商人们希望这些瑞兽能给他们的商业、家庭和婚姻带来好运。

龙在中国文化中有欢腾、喜庆之意,“鱼”的谐音“余”意味着年年有余。此幅《鱼龙漫衍》砖雕表现的是一派喜庆欢乐、生机勃勃的场景,象征着人们追求富足生活的美好愿望。

◉ **A Flying Dragon and Swimming Fish**

This sculpture, in which a dragon is flying in the sky and a flatfish swimming in the water, is called *Yulong Manyan* in Chinese.

In ancient days, *Yulong Manyan* was the name of a acrobatic drama, in which *Yulong*, a kind of legendary gigantic lynx, first performed in a hall, then jumped into the water in front of the palace and turned into a flatfish. The flatfish leaped out and spayed water playfully into the sky to form a mist. Behind the mist, the flatfish then changed into a dragon, which at last soared out of the water and performed in the hall again. Here *Manyan* means " gigantic". *Yulong*, dragon and flatfish were all auspicious animals that were believed to bring good fortune to people. Among them, flatfish is thought to be a kind of fish which always swim in pairs. In ancient times, a flatfish was believed to have its two eyes on one side. Therefore, two flatfishes need to swim together to tell the direction. Later, flatfishes were seen as inseparable people or lovers. It then became a symbol of faithful love in ancient China. Ancient merchants hoped these auspicious animals would bring good fortune to their business and family.

This carving has a vibrant and joyous outlook that conveys the Chinese people's longing for a better life with dragon signifying vigor and fish meaning surplus.

4. 福禄寿三星高照 位置:《参天地》匾额上枋枋心

此图左刻寿星,中刻禄星,右刻福星,故名“福、禄、寿三星”。

古人认为浩渺无际的星空是众神的居所,每一点星光都是一位星辰之神的象征。星辰之神虽然远在天边、遥不可及,却被认为是地上万物的主宰。三星原指参宿三星,心宿三星和河鼓三星。后称传说中的三位神明为"福禄寿三星"。福指幸福、好运,禄象征高官厚禄,寿即长寿。一般认为百岁为上寿,八十为中寿,六十为下寿。三星分别是给人们带来福气和好运的福星,负责加官晋爵的禄星和保佑人们长寿的寿星。

◉ **The Three Stars of Fu, Lu and Shou Shining High**

In this carving, on the left is the God of Longevity, in the middle the God of Official and on the right the God of Blessing.

The ancients considered that the deities occupy the broad sky alive with stars. Each shining star stands for a deity. Though they are faraway, they are considered to be dominators of Earth. At first, they were thought to have horrible appearances like monsters. With the rise of Taoism, they were promoted to star officials. Three of them are of great importance. They are respectively the God of luck who brings people good luck, the God of official who helps people to become government officials and the God of Longevity who grants people a healthy and long life.

5. 夔一足 位置:《参天地》匾额内侧立图

夔,原为古代神话中的一种奇异动物,如龙,长有一足。

根据《吕氏春秋·察传》的记载,鲁哀公问孔子:"舜时的乐官夔是否只有一足?"孔子云:"昔时舜欲以乐传教于天下,乃令重黎举夔于草莽之中而进之,舜以为乐正。重黎又欲求人,舜曰:'若夔者一而足矣,'故曰夔一足,非一足也。""夔"是一个人名,乐正指的是古代司音乐的官员。传说舜想用音乐教化天下百姓,就让重黎从民间举荐了夔,舜任命他做乐正。夔于是校正六律,谐和五声,用音律来调和阴阳之气,因而天下归顺。重黎还想多找些象夔这样的人,舜说:"音乐是天地间的精华,治理国家的关键。只有圣人才能做到和谐,而和谐是音乐的根本。夔能调和音律,从而使天下安定,所以像夔这样的人一个就够了。”所以夔一足指的是“一个夔就足够了”,不是“夔只有一只脚”。后人用“夔一足”或夔的形象,表达有能力和才能的人“一人足也”的意思。

匾额《参天地》两边对称各刻有一幅《夔一足》砖雕,寓意是人与天地并立,人尽其才。音乐是天地之精华,圣人用音乐顺民心、平天下,使得天下太平,和谐昌盛。

◉ **One Kui Is Enough**

Kui is a kind of strange dragon-shaped animal in ancient Chinese mythology, with only one foot. *The Spring and Autumn Annals of lv* records that the Duke Lu Ai asked Confucius, “Did Kui (a Shun-time musical official) have only one foot?” Confucius answered, “Once Shun (an ancient Chinese emperor) asked Chong Li to find a person in charge of music, Chong recommended Kui and Shun assigned Kui head of musical official. Kui then improved music and sound to harmonize *yin* and *yang*, so that Shun was able to subordinate all the people. Chong Li wanted to find another talent like Kui, while Shun said, ‘Kui is so able that one Kui is enough.’ This story was mistakenly passed down as Kui with one foot.”

The two carvings are placed symmetrically on the sides of the plaque read *Can Tian Di*, which sends a message that man is as important as Heaven and Earth, so one’s talents should be fulfilled. And music is the essence of heaven and earth, so by subordinating people with music, the saints can make the world an earthly paradise.

6. 达摩渡江 位置:《参天地》匾额西外侧立柱图

达摩是印度佛教第二十八世祖,于南朝宋末(520至526年间)渡江来中国传播佛教,成为禅宗在中国的始祖。

南朝宋末,达摩从天竺航海至广州,弘扬佛法,以禅法教人。后达摩传法至梁朝都城建邺(今南京)。梁武帝萧衍笃信佛教,闻其名,迎入城内。达摩所传的是禅宗大乘佛法,主张面壁静坐,普渡众生。梁武帝笃信小乘佛教,二人不相契合,达摩决意北上。得知达摩离去,梁武帝试图派人将他追回。而当追兵行至幕府山中段时,两边山峰突然闭合,一行人被夹在两座山峰之间。达摩行至江边,就在江边折了一根芦苇投入江中。芦苇瞬间开出一蓬芦苇花,花分五片,如

船般展开在江面上。达摩踏船而渡，飘然过江。从此这座山峰被称作夹骡峰，而山北麓达摩曾休息的山洞称作达摩洞。北魏孝昌三年(527 年)，达摩到达嵩山少林寺，见此处群山环抱，环境清幽，佛业兴旺，于是在此地落脚开始传教，广集僧徒。此后，少林寺便被认为是中国佛教禅宗祖庭的佛门净土。

此砖雕名为《达摩渡江》，下部刻滔滔江水，达摩足踏一叶芦苇从容渡江，上方是悬崖峭壁，树木繁茂。该砖雕创作是对古诗“一苇渡长江，九年面绝壁；身轻若羽毛，心静恒虚寂”的视觉展现。

◉ **Bodhidharma Cross the Yangtze River**

Bodhidharma, who was supposed to be the twenty-eighth Patriarch in India, came to China some time between 520 and 526, where he became the first Tsu (patriarch, literally, ancestor) of the Chan school in China.

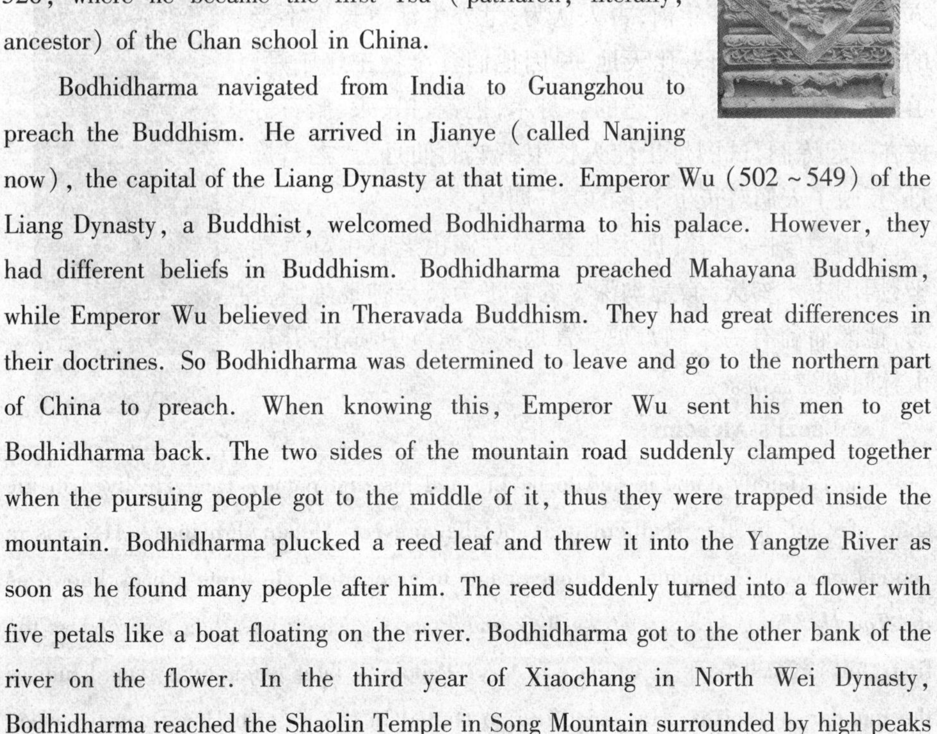

Bodhidharma navigated from India to Guangzhou to preach the Buddhism. He arrived in Jianye (called Nanjing now), the capital of the Liang Dynasty at that time. Emperor Wu (502 ~ 549) of the Liang Dynasty, a Buddhist, welcomed Bodhidharma to his palace. However, they had different beliefs in Buddhism. Bodhidharma preached Mahayana Buddhism, while Emperor Wu believed in Theravada Buddhism. They had great differences in their doctrines. So Bodhidharma was determined to leave and go to the northern part of China to preach. When knowing this, Emperor Wu sent his men to get Bodhidharma back. The two sides of the mountain road suddenly clamped together when the pursuing people got to the middle of it, thus they were trapped inside the mountain. Bodhidharma plucked a reed leaf and threw it into the Yangtze River as soon as he found many people after him. The reed suddenly turned into a flower with five petals like a boat floating on the river. Bodhidharma got to the other bank of the river on the flower. In the third year of Xiaochang in North Wei Dynasty, Bodhidharma reached the Shaolin Temple in Song Mountain surrounded by high peaks and rich forests. He stayed there and started to preach, and took many monks as his disciples. After that, Bodhidharma was regarded as the founder of the Buddism school in China and Shaolin Temple the origin of Buddism.

In this picture, Bodhidharma stands on a reed-leaf flowing on water, which is created according to a poem that describes such a scene.

7. 老君炼丹 位置:《参天地》匾额东外侧立柱图

老君是太上老君的简称,又称老子。他姓李,名耳,字聃,又字伯阳,春秋时楚国苦县人(今河南省南部)。因其著有《道德经》,孔子尝往问礼,后世把他尊为道家始祖。老子被神圣化始于东汉。东汉的张陵(后来的张天师)创立天师道,为了和佛教抗衡便抬出老子为祖师,并尊其为太上老君。

道家为得道成仙、长生不老,视炼丹术为重要的修行方法。早在两千多年前,古人认为,世外生活着长生不老的神仙。这些神仙寿比天地,只因他们食金饮玉。于是,道家创造出许多令人着迷的炼丹术,把铜、铅、汞、硝石等放在一起炼制,试图炼出使人长生不老的仙丹。"老君炼丹"反映了人们对长生不老的美好期望。

砖雕中刻一老者(即太上老君)在深山老林中静静端坐,周围古木参天,寂静幽深。老者正聚精会神地炼制丹药,他的面前有一个制丹炉。青烟从药葫芦中喷出,升至山林间缭绕盘旋。

◉ Laozi's Alchemy

Laozi's family name is said to be Li, and his first name, Tan. He lived in the state of Chu in the southern part of the present Henan Province. He was a contemporary of Confucius, who once came to greet him. He wrote a book known as the *Tao De Ching* (*Classic of the Way and Power*), which was then regarded as the first philosophical book in Chinese history. People of later generations regard him as the founder of Taoism. His sanctification started in the Eastern Han Dynasty, when Zhang Ling founded a new religion called Tianshi Taoism. In order to contend against

Buddhism, he took Laozi as their grandmaster.

As early as two thousand years ago, people thought that there must be an immortal god or goddess who lived in the paradise. They had long life span just because of their nutritious food made of gold and jade. Motivated by this idea, Taoists used copper, lead, mercury and nitrokalite to make elixir which was said to make people immortal like gods or goddesses. Also some people were fascinated by the idea of turning into a millionaire overnight by alchemy, thus so many studies of alchemy came into being. Laozi's alchemy is the reflection of this kind of good wishes to be an immortal man or a millionaire overnight.

In the carving, we see that in the silent mountains, an old man is absorbed in refining elixir, with a smoke rising from the furnace into the forest.

8. 吴越之战 位置:正门牌坊明间大额枋

《吴越之战》砖雕图,来自于中国春秋时代一个真实的历史故事。

春秋吴越之战时,越王勾践先战败于吴王夫差。夫差罚勾践夫妇在自己宫里服劳役来羞辱勾践。越王勾践在吴王夫差面前卑躬屈膝、百般逢迎,骗取了夫差的信任,终于被放回越国。被释放回越国后,勾践卧薪尝胆,不忘雪耻。他采用美人计,挑选了两名绝代佳人西施和郑旦送给夫差,并年年向吴王进献珍奇珠宝。夫差愈发贪恋女色,不再过问政事。丞相伍子胥力谏无效,反被逼自尽。公元前482年,吴国大旱,勾践乘夫差北上会盟之时,突出奇兵伐吴。吴国被越所灭,夫差自刎而死,越王成为春秋时期的最后一位霸主。

此幅砖雕图描绘了吴越两军交战的场景,中间两匹战马上,左为夫差,右为勾践。吴越之战告诉我们,失败并不可怕,只要有决心和毅力,就能像越王勾践那样通过卧薪尝胆,东山再起。这幅作品带给人的思考是深远的。

◉ **The Warfare Between Kingdoms Wu and Yue**

This carving is based on a real historic story of the Spring and Autumn Period.

During the Spring and Autumn Period, Emperor Goujian of Yue was defeated in the first war between Wu and Yue. He was punished by the Emperor Fuchai of Wu, by being a servant in the Wu palace. In order to win Emperor Wu's trust, Emperor Yue pretended to be servile in his presence, which cheated Emperor Wu's allowance of releasing him back to his own country, Yue. Back in Yue, Emperor Yue subjected himself to hardships of all kinds in order to strengthen his resolution to wipe out his humiliation. Meanwhile he succeeded in luring Emperor Wu to indulge in sex and booze. From then on Emperor Wu totally ignored his state affairs. The loyalist minister Wu Zixu tried all he could to persuade him from his degraded life, but in vain. Then Wu Zixu was forced to commit suicide.

The warfare between Kingdom Wu and Kingdom Yue gives us a lesson that we don't need to fear failure, because if we have the courage and determination like Emperor Yue did, we can succeed at last. The story is both enlightening and inspiring to us.

This picture depicts the battle scene between Kingdom Wu and Kingdom Yue, with two men on the horse: Emperor Fuchai of Wu on the left and Emperor Goujian on the right.

9. 全家福·郭子仪上寿　位置:正门牌坊明间小额枋

《郭子仪上寿》的历史故事发生在唐朝。该砖雕图描绘了唐朝名将郭子仪六十大寿时,家人和朝廷官员们为他祝寿的盛况。

郭子仪(697~781年),唐朝大将,化州郑县(今陕西华县)人。以武举累官至太守。因平安禄山有功,升关内河东副元帅、中书令,后封汾阳郡王。德宗继位后尊其为尚父。时值郭子仪六十大寿,七子八婿前来祝寿,热闹非凡,有典故称"七子八婿满床笏"。根据《唐书郭子仪传》记载:"七子八婿,皆显贵朝廷。前去祝寿者,文武百官,七子八婿,唯独郭暧之妻金枝公主未去,后即打金枝。"因郭子仪多子、长寿,符合中国人的人生理想,故“郭子仪上寿”成为各类艺术创作的题材,如雕刻,绘画等。它体现出中国文化对幸福和长寿的向往。

图中郭子仪端坐正堂,身后有一“寿”字清晰可见。堂下文武百官及郭子仪的七子八婿满面笑容前来祝寿。左上雕刻着扶老携幼看热闹的人群,凸显了欢乐的气氛。下额枋刻有象征活泼、欢乐的“水波纹”纹饰,左右两边的斜边饰有“连枝花卉”的纹饰,与图中盛大喜庆的祝寿场面相呼应,构成一幅和谐的画面。

◉ **The whole Family Picture of Guo Ziyi's Birthday Celebration**

This picture shows a grand occasion of Guo Ziyi's 60th birthday in the Tang Dynasty.

Guo Ziyi (697 ~ 781), a general of the Tang Dynasty, was born in Zheng County of Huazhou (Hua County of Shanxi Province for now). He was promoted to a local official after passing military service exams, and then became a duke. This picture shows a scene of bustle and excitement that his seven sons and eight sons-in-law come to offer congratulations to him when his 60th birthday arrives. It is commonly called "seven sons, eight sons-in-law filling the house". All his sons and sons-in-law are officials, confirmed by *Biography of Guo Ziyi*. His daughter-in-law Jin Zhi, also the emperor's daughter, was the only one absent. For this, she was then beaten by her husband once.

In the center of the principal room sits Guo Ziyi, upright; below him are his seven sons, eight sons-in-law and civil and military officials offering congratulations. Upper left are a group of civilians watching, creating a joyous scene.

The carving shows Chinese people's ideal life, so such a theme is often seen on various kinds of art works, which expresses joy, harmony and longevity.

10. 三酸图　位置:《大关帝庙》匾额西侧兜肚

“三酸图”描绘的是中国古代文人的趣闻故事。“三”指三人,“酸”指味道。

宋代文化名人苏轼、黄庭坚、秦观三人应佛印和尚之邀,前往金山寺。佛印欲求其诗词,将备好的桃花醋给三人品尝。三人饮之,诗兴大发,各赋其诗。有诗描述:“相邀金山喜纵谈,风流学士也通禅。闲来去领桃花醋,莫笑吾侪是三酸。”故称此图为“三酸图”。此图表现出三位文人根据其不同的生活体验及价

值观,品桃花醋品出不同的味道。味之酸者以人生为酸,味之苦者以人生为苦,味之甜者以人生为乐。后世根据三酸图又作出《尝醋翁》,描绘分别代表儒释道三教的三人围一醋缸而坐,各自伸指点醋而尝,三人表情各不相同:儒家以为酸,释教以为苦,道家以为甜。从古至今,文人雅士于正统之外,还往往具有率真、贪玩的脾性,不虚伪、不造作。这三个文人雅士,竟然围着一坛桃花醋,酸得龇牙咧嘴,令人忍俊不禁。正如何训田的《春歌》里所唱的:"春有百花秋有月,夏有凉风冬有雪,若无闲事心头挂,便是人生好时光。"

图中右下茶桌前一人为苏轼,紧邻一人即佛印和尚,桌后为黄庭坚,桌左边为秦观。寺院树木环绕,亭台下有一童子在烧水。

◉ **The Different Tastes of Life with Different Experiences**

The carving depicts an anecdote of men of literature in ancient China.

Three eminent figures in Song Dynasty, Su Shi, Huang Tingjian and Qin Guan, were invited by a monk called Fo Yin to Jinshan Temple in Zhenjiang City which is famous for its aromatic vinegar. The monk asked the three of them to taste the vinegar he made. After tasting, they had different expressions. A poem displayed the scene. Each person had a different taste according to their life experience and viewpoint of life. The one who thought it was sour regarded that life was full of sadness; the one who thought it was bitter considered that life was full of sufferings and frustrations; and the one who thought it was sweet viewed life as a joyous journey. Hereafter some people drew another picture of tasting vinegar, depicting three people who respectively stand for Confucianism, Buddhism, Taoism, showing their different attitudes towards life. Confucians think life sour, Buddhists think life bitter and Taoists think life sweet. The picture of three people tasting vinegar proves the saying, "life is full of vicissitude as four seasons; prosperity in spring, and withering in winter. If we hope to have a good life, we should have an optimistic attitude towards life despite of what happened."

In the picture, Su Shi is sitting on the right at the table, with Fo Yin next to him. The two on the left are Huang and Qin. The temple is in deep mountains and surrounded by many trees.

11. 甘露寺·拜乔国老　位置:《大关帝庙》匾额东侧兜肚

砖雕图《甘露寺·拜乔国老》取材于东汉末年三国时期的历史故事。

三国时期,魏、蜀、吴三国鼎立,英杰辈出,演绎出一段波澜壮阔的历史。三国时期许多充满智慧的故事成为了艺术创作的题材。赤壁之战后,刘备借东吴的荆州不还,周瑜向孙权献计,以其妹孙尚香为饵,设下美人计,诱刘备来京口联姻招亲,趁机将其扣为人质,以逼迫其还回荆州。诸葛亮将计就计让刘备前往娶亲。刘备一行到达东吴时,派人到东吴城市南徐采买婚庆礼品,大肆散布刘备入赘东吴的消息,以致全城百姓人人皆知。刘备到东吴后首先拜访乔国老,即孙策和周瑜的岳父、二乔的父亲,叙说特来成亲之事。刘备与孙尚香成亲后,夫妻二人两情欢洽,刘备遂带夫人返回荆州。周瑜在追赶途中,遭遇诸葛亮所设埋伏大败。诸葛亮调侃他"周郎妙计安天下,赔了夫人又折兵"。

这幅砖雕刻的是刘备在镇江北固山甘露寺拜见乔国老的场景。右边太师椅上端坐一威严和蔼的老者为乔国老,左边毕恭毕敬、拱手作揖者为刘备。

◉ **Visiting Zhou Yu's Father-in-law at Ganlu Temple**

This story of the carving happened during the Three Kingdoms Period in ancient China.

The Three Kingdoms Period saw a splendid warring scene, outstanding strategists and geniuses, which were all featured in a number of artworks. Before the War of Chibi, Liu Bei borrowed from Sun Quan the city of Jingzhou in Hubei Province to act in concert with Sun Quan's army to defeat Caocao. But after the war, Liu Bei had no intention of giving it back, so General Zhou Yu offered advice to Sun Quan that they could use Princess Sun Shangxiang (Sun Quan's sister) to lure Liu Bei to come to Wu for marriage, and when Liu Bei came to Wu, they would catch him and force him to give Jingzhou back. Seeing through the purpose of Zhou Yu, Counselor Zhuge Liang decided to beat Zhou Yu at his own game. Thus Liu Bei went to Wu to marry Princess Shangxiang. Upon arriving in city Nanxu in Wu, he immediately sent his men to spread the news that he would marry Princess Sun Shangxiang. The news was so widespread that all the people in the city knew that Liu Bei would marry Princess Sun Shangxiang. With numerous presents, Liu Bei went to visit Senior Qiao, the father-in-law of Sun Ce (the brother of King Sun Quan) and Zhou Yu (the General) in hope that Senior Qiao could act as a matchmaker. The marriage was a success, but Zhou Yu met his Waterloo and became a laughingstock.

The carving shows the scene that Liu Bei is meeting with Senior Qiao in Mountain Beigu of Zhenjiang.

12. 松鹤延年 位置:《大关帝庙》匾额小额枋下侧花板

"松鹤延年"表示吉祥、长寿。

松因其树龄长,经冬不凋,被认为是长寿的象征。这种象征意义为道教所接受,松遂成为道教神话中长生不老的代名词。所以道教认为服食松叶、松根便能飞升成仙、长生不老。松傲霜斗雪、卓尔不群,也象征着高洁的品质。鹤也被道教引入了神仙世界,得道之士骑鹤往返,以鹤为伴。鹤被视为出世之物,也成了超脱的象征。鹤又被称为仙鹤,在民间被视为不死的仙物。

砖雕中刻着八只仙鹤在祥云苍松间飞舞,姿态优美。松、鹤两种仙物同在一图,寓意是祝老人健康长寿。

◉ **Pine and Crane of Longevity**

The carving of *Pine and Crane of Longevity* expresses the wish of luck and longevity.

A pine is often used for implying a long life because it lives long and remains green in winter. This symbolic meaning is accepted by Taoism and pine becomes an important emblem of immortal being in Taoist mythology. So according to Taoism, eating pine leaves and roots can lead to immortality. Being cold-undaunted and unique, it is endowed with nobility. Cranes were also considered as supernatural beings by Taoism, for celestials are often accompanied by them in the stories of Taoism. Cranes were also thought to be immortal animals. The combination of pine and crane conveys the good wishes of longevity.

In the carving there is a vigorous pine-tree, under which a crane is standing erectly. It connotates the best wishes for seniors to be as long-lived as pines and cranes.

13. 魁星点状元　位置:正门牌坊西侧立柱凸出部位

魁星是中国古代神话中“奎星”的俗称。奎星是北斗七星中前四颗星的总称,原是中国古代天文中二十八宿之一,又称奎宿,后被认为是主宰文章的神。

中国艺术形象中的奎星面目狰狞,金身青面,赤发环眼,头上有两只角。明清时代儒者顾炎武《日知录·魁》:“神像不能像‘奎’,而改‘奎’为‘魁’;又不能像‘魁’,而取之字形,为鬼举足而起其斗。”故魁星神像头部像鬼,一脚向后翘起,如魁字的大弯钩;左手持一只墨斗,如魁字中间的“斗”字;右手握一枝大毛笔,称朱笔,用来点定当年中举者的姓名。

作品中刻的魁星的动作为右手高举朱笔,左脚踩赑屃(又名龟趺,是貌似龟的一种祥兽),金鸡独立,意为“独占鳌头”。魁星在中国各地受到文人的祭拜。最初在汉朝纬书《孝经援神契》中有“奎主文章”之说,后世遂建魁星楼、魁星阁以崇祀之。此图边饰为“八吉”纹饰,寓意吉祥。

◉ Kuixing Appointing Zhuangyuan (the Number One Scholar)

Kuixing is one of the 28 constellations in ancient Chinese astronomy, named “Kui constellation”. He is said to be responsible for choosing the top examinees to become officials in Keju system.

Kuixing is artistically described as a god of a horrible looking, with a golden body, a green face, red hair, round eyes and two horns on the head. In the carving,

Kuixing looks like a ghost, with one of his foot raised backward, an ink fountain in his left hand and a red writing brush in his right hand. It is said that he would mark the name of the best examinee (*zhuangyuan*) who will become an official.

Kuixing, with a brush in his right hand, stands on his left foot on the head of a turtle, meaning "the Number One of All Examinees"(独占鳌头). Kuixing was worshiped by Chinese people in old times. They built many temples or towers for him to show their awe to the God who controlled their fate in imperial examinations.

14. 文昌帝君 位置:正门牌坊东侧立柱凸出部位

"文昌帝君"本是星名。他在中国古代传说中是掌管功名禄位之神,又称文曲星,多为读书人所祭拜。

神话传说中,文昌帝君姓张,名亚子,居蜀七曲山(今四川梓潼县北),仕晋战死,被后人立庙纪念,并追封为"英显王";元仁宗延佑三年(1316 年)又被追封为辅元开化文昌司禄宏仁帝君。从此以后,梓潼神张亚子又被称为文昌帝君。随着科举制度的发展,对于文昌帝君的祭拜也逐渐普遍。旧时每年二月初三为文昌帝君神诞之日,官府要员和当地文人学士都要到供奉文昌帝君的庙宇祭祀。

图中文昌帝君立于一白虎头上,周围花团锦簇,环绕着"八吉"的吉祥纹饰,表现出人们对文昌帝君的崇拜敬畏之情,也寓意着对读书人独占鳌头、前程似锦的祝福。

◉ **The God of Wenchang**

Among the Gods adored in folks, the God of Webchang (the God of Literature and Letters), in charge of official rank and salary, was mostly respected by scholars.

According to the legendary saga, the God of literature was originally a man called Zhang Yazi who lived in Mountain Qiqu (in the northern part of Sichuan Province). He sacrificed his life in warfare for the interest of the local people. Later generation built a temple in memory of him. In 1316, he was conferred the title God of Literature in the Yuan Dynasty. With the development of imperial examination, the custom of worshiping him became popular, therefore many temples were built for him. In lunar calender, March the third was regarded as the birthday of God of Literature. On that day, many officials and scholars would worship him in hope of getting his grace by holding a competition of reciting and writing poems in the temples where sacrifices were offered to him.

The God of Wenchang is standing on a tiger head surrounded by flowers. The edge ornament was "the eight lucks", which implies a bright future and a good fortune.

15. 天禄　位置:正门牌坊明间西侧贴角花板

此砖雕所刻神兽名为“天禄”。据中国古籍记载,西域有桃拔、狮子、犀牛三种野兽。桃拔中一角者为天禄,大如斗,形似虎,正黄色,有须髯,尾有绒毛,可攘除灾难,永安百禄。汉代多刻此物为装饰,谓之“天禄书镇”。“天禄”的意思是上天赐予的禄位。《见志诗》云:“富贵有人籍,贫贱无天禄。”

据说天禄神兽能为人们去除灾难、带来财富。作品表达出祈盼得到上天赐予的官位,享受富足生活的愿望。

◉ **Tianlu**

The mythical animal on this sculpture is called *Tianlu*, which is believed to be able to bring good fortune and wealth to people. According to some ancient Chinese books, there were three kinds of rare animals in western China: Taoba, lion, and rhinoceros. Taoba with one horn was also called *Tianlu*, which had pure golden fur, whiskers and a tail with fuzz and looked like a tiger. It was said that *Tianlu* can bring people high status and good salary.

Therefore, ancient merchants carved *Tianlu* to keep themselves from evils and bring them great fortune. So *Tianlu* expresses their wishes for a blessed life.

16. 麟吐玉书 位置:正门牌坊明间东侧贴角花板

麒麟是古籍中记载的一种神物,也是中国传统吉祥兽之一,与凤、龟、龙共称为“四灵”。此兽的外形集龙头、鹿角、马身、鱼鳞、牛尾于一体。在中国文化中,麒麟是一种仁慈、祥瑞的动物。民间有“麒麟送子”之说,并认为麒麟主太平、长寿、福禄。

因此,在中国传统民俗礼仪中,麒麟被制成各种饰物和摆件,用于佩戴和安置家中,有祈福的用意。人们称机灵的小男孩为“麒麟儿”,有称赞其聪明伶俐之意。

◉ **Kylin and Jade Book**

The mythical animal engraved here is Kylin, also referred to as “Qi Lin” or “Lin” in Chinese. It was thought to be one of the Four Divine Animals in ancient China, together with phoenix, turtle and dragon. It was said to have dragon head, deer horn, horse body with scales of fish and ox tail. People in China regarded it as a beast of benevolence and auspiciousness. Folk Chinese generally believe that kylin can bring offspring, peace, longevity and fortune.

Therefore, in Chinese traditional folk culture, kylin often appear in ornaments and decorations to bring people blessings. Sometimes people call clever boys “kylin kids”.

17. 衔环报恩 位置:《大关帝庙》匾额西侧立柱上方竖图

《衔环报恩》取材于中国古代神怪小说。衔环,就是指用嘴叼玉环相赠。“衔环”还常与“结草”连用,作“结草衔环”或“衔环结草”,指报恩。

据古代神怪小说记载,杨宝生于东汉,天性仁慈。他九岁时,在华阴山北看见一只黄雀被鹰追逐咬伤后掉在地上,又遭到蚂蚁撕咬,痛苦挣扎。杨宝怜悯黄雀,将其救回养伤,至创伤痊愈后放飞。当夜杨宝梦见一个黄衣童子向他再三拜谢:“我是西王母使者,在飞往蓬莱仙山途中遭受伤害,承蒙拯救疗养恩德。如今我要返回南海。”说罢,便以四枚白玉环赠送杨宝。又说:“祝愿你子孙清白,位登三公,就如同这玉环一样。”后来,杨宝的仁孝传闻天下,名誉和地位日益显

赫。杨宝的子孙,儿子杨震、孙子杨秉、重孙杨赐和杨彪,果然都位列三公。这是一则关于感恩的神话故事,它告诉我们,人应该慈悲为怀,并且常存感恩之心。

砖雕中刻的人即杨宝,怀抱着一只受伤的黄雀。树上有一只侧首俯视的鹰。中上部刻莲花纹饰,下配万字纹边饰,寓意纯洁高贵,吉祥如意,给人古朴典雅的感觉。

◉ **The Oriole Repaying the Benefactor with Jade Bracelets**

According to ancient fantasy novels, Yang Bao was born with a benevolent nature in the Eastern Han Dynasty (25 ~ 220). At nine, he once saw a yellowbird in the north of Huayin Mountain, which had been pursued and injured by an eagle, fell to the ground. The yellowbird was being bitten by ants and struggling painfully. With deep compassion, Yang took it back and healed its wound. Then he let it fly away. That night Yang Bao dreamed of a boy in yellow who repeatedly thanked him and said, "I am a servant of the Queen Mother of the West. On my way to Penglai Mountains, I was injured. I am grateful for your rescuing me. Now I'll go back to the South Sea." Then the boy presented Yang Bao with four white jade rings and added, "I wish your offsprings could be good and rich people, and pure like the jade rings." Later, Yang Bao's son Yang Zhen, grandson Yang Bing, great grandsons Yang Ci and Yang Biao all became state councilors. This mythological tale tells us that people should always be compassionate and grateful.

On the edge of the carving there are lotus and the shape of "卍" implying fortune and luck. The nude, ugly but good-hearted man is Yang Bao, holding an injured yellowbird, overlooked by a fierce eagle in the tree.

18. 鹿骇狼顾　位置:《大关帝庙》匾额西侧花板(西)

成语"鹿骇狼顾",描述的是鹿和狼各自的习性。鹿性善,易受惊,所以称鹿骇,比喻人们惊惶纷扰的样子;狼行走时,常转过头往后看,以防被袭击,比喻人有后顾之忧。《史记·苏秦列传》载:"秦虽欲深入,则狼顾,恐韩魏之议后也。"意为,秦朝虽欲深入,但有后顾之忧。汉桓宽《盐铁论·险固》:"如此,则中国无

狗吠之警,而边境无鹿骇狼顾之忧矣。”南朝梁陆倕《石阙铭》:“于是治定成功,迩安远肃。欲兹鹿骇,息此狼顾。”此处意为边境安定,国家太平,再无后顾之忧。

本作品刻一惊恐之鹿和一顾盼之狼,中间隔古树对峙相望,气氛紧张。

◉ **Scared Deer and Restless Wolf**

Luhai and Langgu is a Chinese idiom that describes the instincts of the two animals, deer and wolf. Deer have the nature of being vigilant, and wolves like to turn back their heads to see whether there are other assaulting creatures preparing to attack them. When in perilous situation, people are often in cautious state of mind to keep themselves safe. Therefore in Chinese culture, there is a saying that deer are easily frightened and wolves are cautious about the attacks from its back. This saying implies that we should be cautious in any situations in order to keep ourselves safe and sound.

A scared deer and a restless wolf are carved in the picture and they are looking at each other nervously.

19. 心猿意马 位置:《大关帝庙》匾额西侧花板(东)

成语“心猿意马”原是道家用语,字面意思为:心好像猴子的跳跃、像马的奔跑一样控制不住。另一说法是:心猿是佛教用语,比喻攀缘外境,内心浮躁不安仿佛猿猴;而意马则比喻难以控制的心神。该成语的出处很多,如《敦煌变文集·维摩诘经讲经文》记载的“卓定深沉莫测量,心猿意马罢颠狂”;汉代魏伯阳《参同契》的“心猿不定,意马四驰”;语本《维摩经.香积佛品》的“以难化之人,心如猿猴,故以若干种法,制御其心,乃可调伏”;《成神经》则记载,有人在奇怪的荒山上看见一头金毛猿拿着桃子在吃,忽然天上飞过一只天马,金毛猿看见了撒腿就追,于是就有了"心猿意马"的说法。

佛道两家对此语的释义均有佛心不静或道心不定、心不在焉之意,故"心猿意马"比喻人的思绪纷乱,心神不定,难以安静。作品中刻有一匹马在旷野上行走,树下一只猿猴侧身回望。

◉ **When One Means Gibbon, He Thinks of a Horse**

It is a four-character Chinese idiom with negative meaning. It means someone's mind is restless and whimsical. There are many sayings about its origin, but generally speaking, it is related to Buddhism, which refers to those who can't make up their mind or concentrate on what they are doing, like a restless monkey or a galloping horse. The English equivalent might be "when one means gibbon, he thinks of a horse".

The idiom has the meaning that one can not fix his mind on a single thing. In the picture, a horse is walking in the open field and a restless monkey is under a tree not far away.

20. 虎落平原　位置:正门牌坊西边楼大额枋

砖雕《虎落平原》取材于中国古代儿童启蒙书《增广贤文》。该书最早见于明代万历年间的戏曲《牡丹亭》,书中许多谚语充满人生智慧和哲理,成为了警世格言。

成语"虎落平原",全句为"龙游浅水遭虾戏,虎落平阳被犬欺"。平原即平阳。虎原本是山林中百兽之王,然而一旦离开深山,受困于平原,就失去了威风,甚至被狗欺负。比喻有权有势者或有实力的人因为环境的变化失去了自己的权势或优势,就会受到冷落和遗弃。该成语表现出了世态炎凉、人心冷漠,同时也告诫世人:世事难料,所以得势之时不要猖狂。

图刻一只猛虎离开山林落入平原,一群狗正朝它吠叫。它们试图扑向猛虎,但似乎又望而生畏。虎与狗的神情均露出迟疑。虎欲表示友好,而狗却虎视眈眈。

此图以装饰纹样"回纹"饰边。该纹饰呈环状,形如"回"字,寓意"富贵不断头"。

◉ **A Tiger Trapped on the Plain**

It is adapted from a Chinese idiom from a famous ancient book for educating children, *The Collection of Wisdoms*.

A tiger went down from the mountain forest to the plain and a group of barking dogs were trying to pounce on it. Once leaving the mountains, the tiger, the king of beasts, may fall trapped or even be bullied by dogs. So there is a saying that tigers will be bullied on the plain. This sculpture illustrates that bad luck would call on after one loses his power or advantage. The saying gives us a lesson that we should not be too arrogant in our prime, nor should we be depressed in frustration. It is very hard for us to predict what situation we will be in future and no matter what happens we should face it with optimism. The brick carving bears the businessmen's warning that people should keep an awareness of crisis and always unite as one.

In the carving, a tiger and two dogs are glaring at each other on the plain with complex feelings of hatred, hesitation and terror. As we see the carving is trimmed with the edge ornament of the Chinese character "回", which has the message of everlasting prosperity and fortune.

21. 王羲之爱鹅 位置:正门牌坊西边楼额枋之间西侧花板

王羲之(321~379年或303~361年)字逸少,琅琊临沂(今山东)人,东晋著名书法家。官至右军将军,会稽内史。王羲之一生酷爱书法,被后人尊为书圣,其书法博采众长,尤其是行书,被誉为"天下第一"。他的行书《兰亭集序》被认为是行书的绝世之作。他还喜爱观鹅、养鹅。故事"王羲之书成换白鹅"讲述的是他为道士抄写《黄庭经》来换鹅群的故事。《换鹅贴》记载:"右军见山阴道士有群鹅,欲求之,要右军书《黄庭经》以换,右军遂予之。后人以王书《黄庭经》名曰'换鹅贴'。"

作品中,王羲之面前放着书案,案上放有文房四宝;左下角是鹅池,池水清澈,碧波荡漾,一只鹅在水中嬉戏。王羲之边观鹅边挥毫模仿鹅姿写字,写出二十多个不同的"之"字。

◉ **Wang Xizhi's Love of Geese**

Wang Xizhi (321 ~ 379, another record 303 ~ 361), a great calligrapher in the Eastern Jin Dynasty, was from Langxie (a place in nowaday Shandong Province). He was promoted to General Youjun. He is considered "the sage of calligraphy", being one of the most famous calligraphic geniuses in Chinese history. He drew the essence from other's calligraphy and created his own unique style. He was admired by other calligraphers in the following generations. His semi-cursive scripts were especially outstanding. *The Preface to the Lanting Collection of Calligraphy* is considered the greatest masterpiece of Chinese calligraphy in history. He loved raising geese in his life, which became a rich source of inspiration for his calligraphy. There was a story about him. Once he exchanged a piece of his calligraphy for a flock of geese with a Taoist called Shanyin.

In the picture, he sits by the desk on which we can see a writing brush, an inkstone and some paper. On the left hand is a pond with geese swimming in the water. As he is watching the swimming geese, he creates twenty different types of "之".

22. 周敦颐爱莲　位置:正门牌坊西边楼额枋之间东侧花板

周敦颐(1017 ~ 1073),字茂叔,北宋哲学家,道州营道(今湖南)人。道州有濂溪,故周敦颐别名濂溪,世称周濂溪。

周敦颐平生视莲为君子,酷爱秀丽端庄、冰清玉洁的莲花,著有《爱莲说》。他在府署东侧挖池种莲,名为爱莲池,池宽十余丈,中间有一石台,台上有六角亭,两侧有"之"字桥。他盛夏常漫步池畔,欣赏着清香缕缕的莲花。他在《爱莲说》中写道:"予独爱莲之出淤泥而不染,濯清涟而不妖","香远益清,亭亭净植,可远观而不可亵玩焉","莲,花之君子也"。从此莲池闻名遐迩。"说"是古代文体之一,它往往借描绘事物以抒情言志。周敦颐的《爱莲说》正是这种文体中的一篇不可多得的传世佳作。后人多以“出淤泥而不染”来赞美一些人洁身自爱、不媚世俗的高贵品格。

作品下方有一个莲池,正是“爱莲池”。池中莲花盛开,仿佛送来缕缕清香。一座亭子立在池边。濂溪先生正缓步走下台阶观莲。

◉ Zhou Dunyi's Love of Lotus

Zhou Dunyi (1017 ~ 1073) was born in Yingdao, Daozhou (a place in nowaday Hunan Province) where there was a place called Lianxi, so he was also called Zhou Lianxi.

He loved lotus throughout his life, especially the elegance, grace, beauty, sedateness, and purity of lotus. He had people dug a pond beside his residence to grow lotus and called it The Pond of Loving Lotus which is over three meters wide with a stone platform in the middle and a six-cornered kiosk on it. He often wandered around the pond in midsummer to breathe the flagrance of the elegant lotus sent by the wind. In his famous essay, *On the Love of Lot*us, he said, "I do like lotus for it is not imbrued by the mud out of which it grows, nor is it coquettish though bathed in clean water. It is so pure, delicate and bright. It is also straight, proper and honest. It gives a pleasant odor that people can smell even far away. It has no unnecessary branches. It can only be appreciated distantly, without being touched." The fame of Pond of Loving Lotus spread far and wide since then.

The pond in the carving is Zhou's Pond of Loving Lotus, with a few stairs down from the pavilion. The lotuses are in full bloom, you can almost smell their flagrance. Not far away from the pavilion, Zhou Lianxi is taking a walk, enjoying his lotus flowers.

23. 九世同居 位置:正门牌坊西边楼小额枋

此幅砖雕又名“九狮图”、“狮子滚绣球”、“九狮同居”等。图中刻大小九只

狮子,围绕一球在一起嬉戏玩耍,生动有趣。寓意为九代同堂,老人长寿,阖家幸福。

汉语中“狮”的谐音是“世”,“菊”的谐音是“居”。九狮同玩一球,球的形状如同菊花,称为“九狮同菊”,即“九世同居”。在中国,家庭中人丁兴旺、儿孙满堂,几代人同住一个屋檐下和睦相处,被认为是有福气。砖雕《九世(狮)同居(菊)》,寄寓着人们对美好和睦的大家庭的期盼。

砖雕的四周刻有“水波纹”边饰,活泼喜庆。

◉ **Nine Lions Playing the Ball**

There are nine happy lions, big and small, playing together with a chrysanthemum shaped ball in this carving.

In Mandarin Chinese, “lion” and “generation” are both pronounced as “shi”, while “chrysanthemum” and “living” are both pronounced as “ju”. Therefore, this sculpture means “nine generations living together”, which connotes the good wish of having a prosperous and harmonious family. In ancient China, having a large and harmonious family is a great expectation of people. The brick carving bears the expectations of a harmonious family and lots of offsprings.

The carving is edged with ripple ornaments, implying liveliness and happiness.

24. 犀牛望月 位置:正门牌坊西边楼西侧贴角花板

《犀牛望月》图刻祥云环绕瑞树,树下一只犀牛安详地凝视着天空中的月亮。

《犀牛望月》的寓意和文化内涵甚为复杂。中国文化中的很多哲理都体现在神话故事中。一说“犀牛”的谐音是“喜牛”,因此犀牛是象征喜庆的吉祥动物。二说“犀牛望月”源于中国的易学和内丹学说。“牛”是易学中“坤”卦的另一种表述,指大地,对应人的身体;“月”是易学“坎”卦的通俗说法,对应人体的肾精。“犀牛望月”指身心寂静,目光返照,气沉丹田时,肾精化为阳气沿督脉到达头顶的“泥丸宫”(大脑正中)。中国古代哲学认为“一阴一阳之谓道”,阴阳相互作用,此消彼长。而阳从地升,阴从天降。所以通过阴阳交泰(泰卦上坤下乾,指阴阳之交),天地之间始终能保持和谐。犀牛望月也寓意着“天人合一”的期盼。

◉ **A Rhino Watching the Moon**

The sculpture is known as *A Rhino Watching the Moon*. In the sculpture, a rhino is staring at the moon peacefully under the tree.

The essence and philosophical theories of Chinese culture are sometimes expressed by myths. One saying goes that rhino is a good animal from heaven, which has the same pronunciation with *xiniu* in Chinese, meaning an animal that will bring good fortune. There is also another saying that the story *A Rhino Watching the Moon* originates from *Yi Jing* and *Inner Alchemy Theory*. Ancient Chinese philosophy stressed on the harmony between man and nature. It is believed that everything in the universe has two opposite elements, *yin* and *yang*, and with the two exchanging alternately, there can be harmony between Heaven and Earth, and between universe and man. This is the principle of the two opposite sides in nature. Chinese culture regards human life as a part of nature so the only way for us to survive is to live in harmony with nature. Thus *A Rhino Watching the Moon* implies Chinese people's wish for the harmony between man and nature.

25. 万象更新 位置:正门牌坊西边楼东侧贴角花板

“万象更新”指事物或景象焕然一新,大地呈现一派新气象。

古人以象为吉祥物,中国古代南方诸国历代遣使进献驯象者。《魏书》卷十二载,“元象元年(538 年)正月,有巨象自至砀郡陂中,南兖州获送于邺。丁卯,大赦,改元。”另据文献记载,两宋时期,宫廷中设有象院。每逢明堂大祀,都有象车游行。北宋时的汴京和南宋时的行都在明堂大祀时“游人嬉集,观者如织”,“土木粉捏小象儿并纸画,看人携归,以为献遗”,“外郡人市去,为土宜遗送”。中国以农历正月初一为一年的开始,固有春联“一元复始,万象更新”。

在雕刻、绘画等艺术表现形式中,"万象更新"的涵意常常以象的背上驮一盆万年青,或象的披巾上的"万"字(有万福和万寿之意)来表达。此图刻一只悠然自得的大象,观察着周围生机盎然的景象。"大象"和"景象"的"象"是同一字,故象寓意着世间万事万物生机勃勃,焕然一新。

◉ **Everything in a New Aura**

The Chinese idiom meaning everything in a new aura contains our hope that everything in our lives can be renewed.

In traditional Chinese culture, elephant was regarded as an auspicious animal. In the Song Dynasty, when people worshipped Heaven, they would have elephants draw carts and parade in the street. According to the historical record of the Song Dynasty, Kaifeng (the capital at that time) was crowded with visitors who came for trade or for the parade. In Chinese lunar calendar, the first day of January is regarded as the beginning of the year, as a couplet goes, "At the first day of the year, everything is in new aura."

The image that everything takes on a new look is shown by an elephant with some evergreen plants on its back, or a pattern of 卍 (symbolizing good luck and happiness) on its shawl. The carving shows an elephant standing in the open air, looking at the vigorous scene curiously. In Mandarin Chinese, the character "象" *xiang* has the meaning of both "elephant" and "things". The brick carving implies that all things have a new look and look fresh and wonderful.

26. 六合同春 位置:《大关帝庙》匾额东立柱上方竖图

"六合同春"为吉祥词语,意思是大地同春、安泰祥和。六合指天地和四方,亦泛指天下或宇宙。李白《古风》诗:"秦王扫六合,虎视何雄哉!""扫六合"即"得天下"之意。

"六合同春",又称"鹿鹤同春",在明代也有以"六鹤同春"来表现的。杨慎《升庵外集》卷九十四:"北之语合鹤迥然不分,故有绘六鹤及椿树为图者,取六合同春之义。"民间运用谐音的手法,以"鹿"取"六"之音,"鹤"取"合"之音;"春"的寓意则用花卉、松树、椿树等来表现。这些形象组合起来,构成了"鹿鹤同春"的吉祥图案,其谐音为"六合同春"。

作品中刻一株苍劲古松,一猿猴正攀援而上。松下两鹿似在幽幽和鸣。这幅松鹿图砖雕寓意着春天来临,万物欣欣向荣。底边刻有万字纹寓意吉祥,上边刻莲花边饰寓意高贵纯洁。

◉ **The Whole World in Spring**

Spring time symbolizes prosperity. In Chinese culture, *Liuhe* refers to the world. Because *Liuhe* sounds just like deer and crane in Chinese, people use the two animals to stand for it. *Liuhe tongchun* means that the world is in spring and everything is in prosperity. People often use deer in Chinese to refer to the land, and crane to symbolize harmony because they are homophones. And spring is represented by flowers and trees. The images are combined to show the vigor in spring.

This sculpture presents a scene of an old pine tree in spring with a climbing ape on it and two bleating deer under it. It has the message that when spring is coming, everything becomes prosperous. As we can see that at the bottom of the artwork, the edge ornament formed by "卍" resembles luck, while at the upper, the ornament lotus stands for nobility and purity.

27. 狻猊 位置:《大关帝庙》匾额东侧花板(东)

传说中的神兽狻猊为龙所生的九子之一。古书记载狻猊是与狮子同类,能食虎豹的猛兽。《穆天子传》:"狻猊野马,走五百里。"

砖雕中刻一狻猊,其形如狮。古人认为狻猊能带来好运,其多见于中国古建筑装饰和铜铁器艺术品纹饰中。

◉ ***Suanni***

The mythical animal like a lion carved on sculpture is called *Suanni*, which is said to be one of the nine legendary sons of dragon. Ancient books recorded that *Suanni* was a kind of mighty looking similar to the lion which could eat other beasts like tiger or leopard. It was powerful enough to lead and command other animals. As *Suanni* can bring people good luck, it is commonly used in the decoration of ancient Chinese architecture.

28. 怒蟾斗狮 位置:《大关帝庙》匾额东侧花板(西)

"怒蟾斗狮"是一个寓言故事,蟾即蛙。元好问《蟾池》诗:"小蟾徐行腹如鼓,大蟾张颐怒如虎。"《韩非子·内诸说上》:"越王勾践见怒蛙而式之。御者问:'何为式?'王曰:'蛙有气如此,可无为式乎?'士人闻之曰:'蛙有气,王犹为式,况士人有勇气乎?'"勾践为雪战败之耻,见到一只鼓足气的青蛙,便向青蛙行礼表示尊重,以求得勇士。后比喻在强者面前不甘示弱,鼓足勇气与之斗争之意。

砖雕中刻一怒蟾,面对兽中之王狮子,与之对视而毫无惧色,英勇无畏。在此"怒"指的是奋发向上的精神状态,勇往直前的拼劲和不服输的顽强意志。

◉ **The Fury Toad Fighting the Lion**

In the carving there is an angry toad facing the lion. The fury of the toad displays its ambition to beat the lion and its unwillingness to surrender, showing great courage. Sometimes it is necessary for us to have a valiant spirit when facing with a strong competitor. No matter how mighty the competitor is, we should have the courage to face him and the confidence to win.

In the carving, an angry toad is facing a fierce and powerful lion without any fear. It shows its daring spirit, determination and indefectible courage.

29. 三阳开泰 位置:正门牌坊东边楼大额枋

"三阳开泰"的典故来自《易经》。正月为阳,天地交泰。"泰"是《易经》中的一个召福卦象,"所以得名为泰者,止由天地气交而生养万物。"《易经》:"十月为坤卦,纯阴之象。十一月为复卦,一阳生于下;十二月为临卦,二阳生于下;正月为泰卦,三阳生于下。冬去春来,阴消阳长,有吉祥之象。"故以"三阳开泰"象征一年的开端。有一副对联的内容是"三阳开泰,五世其昌"。羊字古时同祥字,寓意为吉祥。因"羊"与"阳"同音,三羊即"三阳",即初九、九二、九三,因为一年中的这三天阳气盛极而阴气衰微。开泰即开启。因此,"三阳开泰"代表着开启新的一年,是岁首吉祥的象征,有世代昌盛的意思。

此砖雕位于大关帝庙正门左侧上方,砖雕四周饰有寓意吉祥富贵不断头的"回纹"装饰。图刻三只羊、一匹马,中间隔着一座桥,呈现出一派生机盎然的景象,象征着旧的一年(马年)已经过去,新的一年(羊年)悄然到来,万物复苏,吉祥如意。

◉ **Three Yangs Bring Auspices**

The saying *Three Yangs Bring Auspices* came from *the Book of Changes*. It means the departure of winter and the arrival of spring, symbolizing a propitious omen, so it is often referred to as the beginning of a year. One couplets read, "Three sheep bring auspices, five lions live together and prosper." In Chinese philosophy, *yin* and *yang* are the two basic elements in nature, the former feminine and negative, the latter masculine and positive. The two elements act alternately so that the universe keeps its balance. The Chinese equivalent of the word sheep is of the same pronunciation as *yang*, so sheep here refers to *yang*.

The carving shows three sheep and a horse, between them is a bridge. It presents a scene of spring full of vigor, and symbolizes a new year of hope and luck is coming.

Additionally, the edging ornament of Chinese character "回" conveys the message of everlasting prosperity and fortune.

30. 鲁隐公观鱼 位置:正门牌坊东边楼额枋之间西侧花板

"鲁隐公观鱼"的故事出自我国古代第一部叙事详细的编年史著作《左传·隐公五年》一书中的《臧僖伯谏观鱼》。它记述了鲁隐公任帝第五年春,臧僖伯谏鲁隐公"如棠观鱼"的故事。

鲁隐公名字里的"隐"字意为隐忍,这个字的由来涉及嫡庶制。中国古代实行一夫多妻制,但各个妻子的地位不平等,这种差别就是嫡庶之分。嫡出指正妻所生子女,庶出则指姬妾所生子女。鲁隐公是鲁惠公的庶出长子。成年后,父亲鲁惠公准备让他娶宋国少女仲子为妻。然而仲子到了鲁国后,父亲惠公见她美丽,便自纳之并立为正妻。不久仲子为惠公生下公子允,允被立为太子。按周朝的传统礼法,立嫡以长不以贤,立子以贵不以长。惠公死时,太子允(即鲁桓公)还小,于是根据惠公的遗命,隐公上台执政,但不是正式继位,而是摄政。《春秋经》云:"不书即位,摄也。"于是,后人赐予他的谥号为隐公。

隐公执政的第五年春,隐公不顾大夫臧僖伯苦口婆心的劝阻,执意要到边境棠地(鲁地名,近宋鲁边界,今山东省鱼台县附近)去看捕鱼。按当时的礼法,打鱼是贱业,身为诸侯王去看捕鱼是"非礼也"。所以《春秋》曰:"公矢鱼于棠",有讥讽之意。

这幅砖雕作品左下角刻一渔人打鱼,右一人即鲁隐公,在俯身观看捕鱼。

◉ Duke Yin of Lu Watching Fishing

This story is based on the episode *Zang Xibo Admonished Duke Yin against*

Watching Fishing in *Zuozhuan*.

There is a story about the name of Duke Yin of Lu. *Yin* means being humble and low-profile. This name had something to do with the wife-concubine system back in old times. Polygamy was the custom of ancient China, while the wives of one man didn't have equal status. Only one of them was considered the principal wife whose sons had the exclusive right to inherit the father's title and fortune. Other wives were called concubines, whose sons had no access to the father's fortune. Duck Yin of Lu was the oldest son of his father, Duke Hui of Lu. He was born by a concubine. When he reached adulthood, his father planned to have him marry a girl named Zhongzi from Song State. However, when the girl reached the Lu State, his father found her beautiful and made her his own principal wife. Then she gave birth to a son named Yun who was then conferred the successor of Duke Hui of Lu. According to the law of the Zhou Dynasty, the king's oldest son born by his principal wife would be his successor. When Huigong died, Yun (i. e. Lu huangong) was still too young, therefore according to his father's last words, Duke Yin came in power as a regent while his younger brother had the title of king. Yin lived a low-profile and humble life, as if he was invisible. *Doctrine of Spring and Autumn* recorded, "not the king, but a regent." Hence the name was given to him after his death.

In the fifth spring after he took power, he went a long way to Tangdi (a place in the state of Lu) to watch fishing in spite of the opposition from his councilors. According to the customs then, fishing was a menial profession. It was very inappropriate for a ruler to watch fishing. So *Spring and Autumn* recorded the story ironically. In the picture, Duke Yin was watching fishing attentively.

31. 陶渊明爱菊 位置:正门牌坊东边楼额枋之间东侧花板

陶渊明(365 或 376 ~427 年)名潜,字渊明,寻阳柴桑人,人称靖节先生,东晋著名诗人。曾任江州祭酒、镇军参军、彭泽令。陶渊明志趣高洁,不慕名利,由于不满现实而决心归隐。靖节先生平生独爱菊,因为菊在深秋开放,此时百花凋零,而菊独抱幽芳,高贵不凡。在陶渊明的诗文中,描述菊的语句很多。如"采菊东篱下,悠然见南山","三径就荒,松菊犹存",足见其爱菊之甚。

该砖雕下刻一园秋菊傲霜盛开，陶渊明手捧一束菊花，居高临下，凭栏而观。

◉ **Tao Yuanming's Love of Chrysanthemum**

Tao Yuanming (365 or 376 ~ 427), a great poet of the East Jin Dynasty, was born in Caisang, Xunyang. His given name was Qian. He was also named Yuanliang, Yuanming, or Jingjie. He was once a high-level official, with noble aspiration and little interest in fame and wealth. He decided to quit his job and return to the countryside for his dissatisfaction with the society. He loved chrysanthemum exclusively throughout his life, because they blossomed in late autumn alone and lived in seclusion when other flowers withered. He mentioned them a lot in his poets. For example, "Pick a chrysanthemum near the east fence, and leisurely I see the mountain in the south" in *Drinking*, and "Though the paths in the garden have nearly been wasted, the pine trees and the chrysanthemums are still there" in *On Returning*. These poems proved his love for chrysanthemums.

This carving shows a garden of blooming chrysanthemums in a late-autumn day. Yuanming is admiring them above, leaning on the railing. Tao Yuanming loved chrysanthemum mainly because it had similar personality with his, for it stood for the hermit in flowers.

32. 五世其昌 位置：正门牌坊东边楼小额枋

这幅砖雕的左右两侧刻着缠绕丝带的两扎书信，中间刻着两大三小共五只狮子在戏耍绣球，活泼可爱，其乐融融。图的四周刻有水纹边饰，有活泼喜庆的寓意。

狮子是百兽之王，同时也被古代中国人认为是神兽，能给人带来吉祥，去除灾祸。在中国装饰文化中,小狮子活泼可爱、大狮子威武高大。在普通话中,“狮”与“世”谐音,此处五只狮子寓意“五世”,本幅砖雕因而得名《五世其昌》,表达出中国人对家族昌盛,子孙延绵的愿望,体现了中国文化浓厚的宗族观念。

◉ **Five Lions Playing the Ball**

Carved on this sculpture are two big lions (one male and the other female) and three small lions playing with a silk ball together. The sculpture is trimmed with ripples and fine woven patterns, beautiful and dignified, merry and cheerful.

Lions were thought to be a traditional mythical animal in China that were believed to be able to bring good luck and drive away scourge. In Mandarin Chinese, “狮” (lion) and “世” (generation) are pronounced in a similar way. Here five lions simplify “five generations” and the sculpture is therefore called “prosperity for five generations in the family”.

33. 一品当朝 位置:正门牌坊东边楼西侧贴角花板

“品”表示中国古代官吏的等级,“一品”为封建王朝官员的最高等级。古人也用“品”来评定动物的等级,最初的“一品”指的是鹤。鹤性清高,为鸟类之长,常被用作一品官补子的图案,故用鹤表示一品。寓意“一品当朝”的艺术作品中,往往有一仙鹤立于惊涛拍岸的海边的礁石上。“潮”与“朝”谐音,故“当潮”暗指“当朝”,即参与政事。

此砖雕刻一头大面阔的雄狮张开大口站在海边。因为狮是百兽之王,群兽对其十分敬畏,也称其为“一品”。明清瓷器也常见此纹饰,是一品官员的标志。常用“一品当朝”来形容身居要职的官员。

◉ **Number One in the Imperial Court**

This carving is called *Number One in the Imperial Court*. *Pin* is the ancient official rank, which ranges from *Pin* One to *Pin* Nine with *Pin* One the highest. When people were trying to rank the animals, they made crane the first *Pin* animal. The crane was often painted standing on a rock near the sea. In Chinese, tide and court are pronounced the same way. Thus, *Number One in the Imperial Court* implies a high position in the court. Similarly, the lion is the king of beasts and is revered by other beasts. Hence the sculpture got the name above mentioned.

34. 唐　位置:正门牌坊东边楼东侧贴角花板

“唐”是中国古代传说中的怪兽,形似麒麟。独角为麟,双角为唐。唐的皮很坚韧,刀剑不能破,传说古人常用其皮做铠甲。在中国文化中,“唐”是一种瑞兽,能为人们带来吉祥。故中国建筑装饰和艺术品中常见这种类似麒麟的怪兽。

砖雕中刻一凶兽,两角竖立,皮似铠甲,即为“唐”。

◉ **The Legendary Animal Tang**

Tang is a legendary animal which looks like Kylin. The difference lies in that Kylin has one horn while Tang has two. Tang's skin was said to be so tough that even a sword can not thrust into it. So its skin was often used to make armor in ancient times. Chinese people believed that Tang can bring good fortune to them.

The fierce animal in the picture with two sharp horns is called “Tang”, which was often seen in artworks, such as architectural ornaments.

35. 二龙戏珠　位置:正门牌坊明间拱门上方

龙是一种神物,而珠是珍贵之物。“单龙戏珠”和“双龙戏珠”的图案常见于建筑装饰以及服饰中。

传说中的龙珠指夜明珠或珍珠,可避水火。《庄子》云:“千金之珠,必在九重之渊,而骊龙颔下。”《埤雅》也提到“龙珠在颔”。中国民间传说龙能降雨,一旦遇到旱年,人们往往就祭拜龙王祈雨。后来产生了名为“耍龙灯”的民俗活动,其目的是祈雨或庆祝丰收。

关帝庙正门呈拱形的门头上雕刻着两条左右对称的龙,正中一颗宝珠闪闪发光,与钟楼门头上的“龙凤呈祥”图案相对应。该砖雕位于关帝庙的正门,象征祥瑞、和谐、高贵和吉祥。

◉ **Two Dragons Playing with a Pearl**

Two Dragons Playing with a Pearl is carved on the arch gate of Guan Di Temple with a glaring pearl. Such a picture with one or two dragons can often be seen as ornaments of architecture or clothes.

Zhuangzi read, “A precious pearl hides in the deep sea and only a dragon can get it.” In Chinese folktale, a dragon can generate wind and rain, so people would pray to the dragon for rain so that crops can grow. The ritual of praying then developed into the activity of “playing dragons” to celebrate the harvest.

Dragon is thought to be one of the divine animals and pearl is valuable, so when they are put together, it means something very precious. The carving also indicates luck, harmony, nobility and good will.

钟楼牌坊

Carvings on the Gate of Bell Tower

36. 松鹤延年 位置:钟楼牌坊上层花板

松与鹤分别是自然界常见的植物和动物。而在古人的观念里,它们则被赋予了各自的象征意义。《松鹤延年》是根据松与鹤各自的象征意义创作的。

据古书《神境记》载,松四季长青,有“百木之长”之称。据说寿过千年的松树分泌的松脂会变成茯苓,服者可得长生。故古人学道,爱择古松下为修炼处,为的就是食用茯苓助力。鹤在中国文化里是长寿鸟,被称为“百羽之宗”。在汉语的敬语里,老人的年龄被尊称为“鹤龄”。

将《松鹤延年》图赠予高龄夫妇,可表达祝两位老人健康长寿的愿望。此图将松与鹤放在一起,是依据“千岁之鹤依千年之松”的说法,意在祝福老人能像松与鹤那样长寿。

◉ **Pine and Crane of Longevity**

Pines and cranes are commonly seen in nature, while in Chinese culture they have a special indication of longevity, for pines remain green in winter and cranes are thought to have a long life. These symbolic meanings were accepted by Taoism, thus the two became important prototypes of immortal being and an emblem of immortality in Taoist mythology. So it was said that eating pine leaves and roots could lead to immortality.

Being cold-undaunted and unique, pines were endowed with nobility. Cranes were also considered as a supernatural being by Taoism, for celestials are often accompanied by them in the stories of Taoism. Cranes were also thought to be immortal animals. People often refer to the age of an elderly person as “crane age” to show their respect. The combination of pine and crane conveys the wishes of longevity.

37. 大梁城 · 范雎逃秦　位置:钟楼牌坊大额枋

“范雎逃秦”是《战国策·秦策三》所记载的一个真实的历史故事。大梁是战国时魏国的都城(今河南开封市西北)。范雎,字叔,战国时期魏国人,著名辩士,秦国一代名相,是我国古代在政治、外交等方面极有建树的政治家、军事谋略家。

战国时期是一个纵横捭阖的时代,出现了许多杰出的辩士和谋略家。范雎善辩,本欲求官于魏王,一展平生所学,但因家贫没有门路,便到中大夫须贾门下当宾客。《战国策》记:“范雎因事从中大夫须贾使齐,齐襄王闻雎贤,赠以厚礼。须贾疑雎以阴事告齐,贾归告魏相,魏齐(相名)使舍人笞击雎,折肋擢齿。雎佯死,箦卷投厕中,守者郑安平,诈言弃尸,阴纵放之。后雎化名张禄,由王稽、郑安平帮助逃往秦国。秦昭王四十一年(前226年)奉任秦相,封于应(今河南宝丰县西南),世称应候。”

作品刻的是范雎逃亡秦国的场景,车上右边的人是范雎,另两人是王稽、郑安平。

◉ **Fan Sui Fled from Daliang City to the Qin State**

The carving is based on a real story in *Strategies of the Warring States*, a book of documentary literature. Daliang city was the capital of the Wei State in the era of the Warring States. It located to the northwest of nowaday Kaifeng city in Henan Province. Fan Sui, a native of the Wei state in the Warring States, was regarded as one of the most prominent statesman and diplomat in ancient China.

Fan Sui, eloquent and witted in diplomacy, had worked for Xu Jia (an officer). When they served as envoys to the Qi state, King Xiang of Qi appreciated Fan's talents and presented him with a lot of gifts. Xu Jia suspected that Fan revealed secrets of Wei to Qi. After their return, he told his doubt to the Prime Minister Wei Qi and the latter had Fan whipped until Fan's ribs and teeth were broken. Pretending to be dead, Fan was put into a basket in the toilet. The jailor Zheng Anping lied that the corpse was deserted, and set him free secretly. Later, Fan fled to the Qin state with the help of Wang Ji and Zheng Anping under the alias of Zhang Lu. In the 41st year of King Zhao's rule of the Qi state, he was widely respected and given a title of prime minister. His fief was in Ying (a place to the southwest of nowaday Baofeng County, Henan Province) and his family was given the hereditary title of Marquis.

This picture presents the scene of his escape to the Qin state. On the chariot, the man on the right is Fan Sui, the other two are Wang Ji and Zheng Anping.

38. 白蛇传　位置:钟楼牌坊小额枋

《白蛇传》是中国传统戏剧曲目,取材于民间神话传说故事。

故事内容是修炼成仙的蛇妖白素贞与青年许仙之间发生的爱情故事。白蛇(又名白素贞、白娘子)思凡下山时,与侍女青蛇(小青)同至杭州。白娘子与店伙计许仙产生爱情,结为夫妻。而法海禅师以白蛇是妖为由,几次试图从中破坏这对夫妇的感情。端午节时,法海要许仙请白娘子饮雄黄酒,酒后白娘子现出白蛇原形,差点吓死许仙。白娘子历尽千辛万苦盗来仙草(灵芝),救活了许仙。而许仙被法海骗至金山寺出家。随后法海借佛法将白娘子镇于杭州的雷峰塔下。

砖雕左边为雷峰塔,塔前站立者即白娘子,右为法海和尚,面前下跪者乃许仙,中为杭州断桥。右上方有一扇庙门,这座庙就是金山寺。图中人物形象生动,山景立体逼真。

◉ **Madam White Snake**

Madam White Snake is a traditional Chinese drama based on a folktale. It is about the white snake (Bai Suzhen, or Madam White) yearning for the mortal world.

The drama tells a love story between Madam White Snake, a fairy, and an ordinary young man named Xu Xian. When the fairy Madam White Snake went down from the mountain with her maid Xiaoqing, she fell in love with a waiter in the inn she lodged, Xun Xian and got married with him. The master of Buddism Fa Hai thought that the white snake was a monster and tried to destroy the marriage several times. On the Dragon Boat Festival, he asked Xu Xian to invite Madam White Snake to drink realgar wine. After she drank the wine, she turned into a snake, which she used to be. Xu Xian was almost frightened to death. Madam White Snake suffered a lot and even stole the ganoderma to save Xu Xian. After Xu's recovery, he was cheated by Fa Hai and captured in Jinshan Temple. Later Fa Hai imprisoned the white snake under the Lei Feng Tower by the power of Buddha.

On the left of the picture is Lei Feng Tower, before it is Madam White Snake. On the right is the monk Fa Hai with Xu Xian kneeling before him, in the middle is Duan Bridge. On the right side of the carving, some stairs lead to an open gate of the Jinshan Temple, where Xu Xian went to be a monk.

39. 蟠桃孝母 位置:匾额《钟楼》西侧兜肚

"蟠桃孝母"的故事来自于中国民间神话传说。

据说有一白猿(猿猴所变)很孝顺。一日,其母久病初愈,想吃蟠桃(仙桃)。白猿买不到蟠桃,就到孙膑(战国著名的兵法家,孙武后代,著有《孙子兵法》)家的桃树下去偷,不料被孙膑发现,捉住绑起。白猿跪泣道:"我并不是有意要偷桃子,只是我母亲年事已高,久病初愈想吃桃子,我买不到只好来偷。"孙膑听了,认为白猿是个孝子,就放了它并且把蟠桃赐给了它。后世常用这个典故形容有孝心的人或尽孝的行为。

在中国文化中,蟠桃被视为仙桃、寿桃,其来历与孙膑为母祝寿有关。相传孙膑18岁离家,到千里之外的云蒙山向鬼谷子学习兵法,一去就是12年。一日他突然想起老母的80岁生日,遂向师父请假,回去为母祝寿。师父送桃与孙膑让他带回给母亲祝寿。孙膑到家后,向母亲献上寿桃。谁知桃子还没吃完,母亲的容颜就变得年轻起来。此后为让老人健康长寿,人们便效仿孙膑,在老人生日时献仙桃祝寿。传说这也是白猿去偷孙膑家蟠桃的缘由。

作品中刻有一白猿手捧所赐仙桃,正举步过桥,准备回家孝敬母亲。左上方两人为孙膑及家人。

◉ **An Ape Steals Pantao for His Mother**

The legend is about a filial white ape. One day his mother just recovered from an illness and wanted to eat Pantao (a kind of precious peach which is said to make people younger). In order to meet her need it decided to steal the peaches from Sun Bin's (militarist, Sun Wu's descendant, author of *Sunzi' Art of Wars*) peach garden. Unfortunately he was caught by Sun Bin. He knelt down and said tearfully,

"Necessity knows no law. My mother has just recovered from an illness and wants to eat *pantao*. I cannot find it in the market, the only way to get them is to steal them from your house." After hearing this, Sun Bin thought the ape was a filial son. He set him free, and presented him with the peaches for his mother.

In Chinese culture, peaches are regarded as a kind of fruit which can make people live long and healthily. So people presented the elderly with peaches as gifts on their birthdays, to convey the wishes for their longevity. The carving shows that the ape is going home with the peaches, and on the left stand Sun Bin and his servant.

40. 蒸梨休妻　位置:匾额《钟楼》东侧兜肚

"蒸梨休妻"是曾子的故事。曾子(前505~前436年)名参,原名曾黎,字子舆,春秋末鲁国人,为孔子四大弟子之一,以孝著称。十六岁拜孔子为师,勤奋好学,深得孔子真传。曾子著述有《大学》、《孝经》等儒家经典,提出"吾日三省吾身"(《论语·学而》)作为修炼的方法。他是孔子学说的主要继承人和传播者,在儒家文化中具有承上启下的重要地位,后世儒家尊他为"宗圣"。

传说曾子家住着他的老母和妻子。一次妻子为母亲蒸梨,蒸得不熟,母亲不愿意吃。曾子对母亲十分孝敬,就因此把妻子休了。《白虎通谏诤》传云:"曾子去(休)妻,梨蒸不熟。妻问曰:'妇有七出:无子,一也;淫也,二也;不事舅姑,三也;口舌,四也;盗窃,五也;妒忌,六也;恶疾,七也。不蒸亦预乎?'曾子曰:'吾闻之也,绝交令可友,弃妇令可嫁也,蒸梨不熟而已,何问其故?'""妇有七出",即休妻的条件有七个。曾子本名曾黎,后因"蒸梨"与曾黎读音相似,这个故事就被叫做蒸梨(曾黎)休妻。

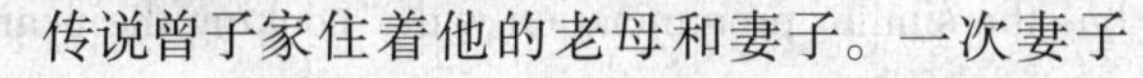

曾子《孝经》曰,"夫孝者,天下之大经也。夫孝,置之而塞于天地,衡之而衡于四海,施诸后世而无朝夕","慎终追远,民德归厚矣"。就是说,孝是天下最主要的法则,把它立置便顶天立地,把它横放便横盖四海,把它延续到后世便会永久存在。所以要谨慎地办理父母的丧事,虔诚地祭祀、追念祖先,这样百姓就淳朴厚道了。曾子的重孝思想和对孝的身体力行为世人称道,被后世列入中国“二十四孝”之一。

砖雕中左边两人是曾子夫妇,右立一老妪就是曾母。

◉ **Zeng Li Divorces His Wife**

This story is about Zengzi (505B. C. ~436B. C.). His given name was Can, and other names for him were Zeng Ziyu and Zeng Li. He was born in the Lu State in late Spring and Autumn Period. He was famous for being a filial son and one of the disciples of Confucius. With his works, *The Schools* and *The Book of Filial Piety*, he was regarded as the successor and one of the prophets of Confucianism.

It was said that Zeng Li had an old mother and a good wife. Once upon a time, his wife steamed some pears for his mother. For the pears were not fully cooked, his mother refused to eat them. At hearing this, Zeng Li divorced his wife angrily to show his filial piety. Knowing that Zengzi wanted to abandon her just because of the uncooked pear, his wife argued with him, “There are seven conditions for divorcing a wife: the first is not bearing a child; the second, not loyal to the husband; the third, not filial to in-laws; the fourth, talkative; the fifth, envious; the sixth, theft; the seventh, a malignant disease. Which one did steaming pears violate?” Zeng Li answered, “Breaking off relations enables one to make new friends, divorcing makes one marry again. There is no doubt for a husband’s decision.” Interestingly, in Chinese, the name of Zeng Li has the similar pronunciation with “cooking the pear”, hence *zheng li*.

Zengzi thought filial piety was the basic and vital one among all virtues. Everyone should respect his parents while they are alive and give them a decent funeral when they depart. Thus the civilians could be of high morality.

As we can see in the carving, the two people on the left are Zeng Li and his wife; beside them is Zeng’s mother.

41. 传胪赐宴 位置:钟楼牌坊拱门上方西侧贴角花板

“传胪赐宴”的说法源自于中国古代的科举考试。传胪是中国古代殿试结果揭晓后唱名的一种仪式。赐宴,指宴请。

中国古代科举考试有乡试、会试、殿试三种。殿试又称御试、廷试，是由皇帝亲自主持的最高级别的考试。明、清殿试中，考生中举后，按成绩分为三甲。第一甲共三人，第一名称“状元”，第二名称“榜眼”，第三名称“探花”，皆赐进士及第；第二甲若干名，赐进士出身；第三甲若干名，赐同进士出身。到了清朝则称二甲的第一名为“传胪”。“传胪”就是依次唱名，传呼新进士进殿觐见皇帝。殿试结果公布之日，皇帝至殿宣布名次，由阁门承接，传于阶下，卫士六七人齐声高呼其名，谓之传胪。“传胪赐宴”，即殿试结束，皇帝宣布登第的进士名次的典礼之后，宴请新进士的仪式。这种制度始于宋代。宋代太平二年，宋太宗亲自策问吕蒙正以下，并赐进士及第，赐宴开宝寺。“传胪赐宴”由此开始。

此砖雕作品刻芦雁三只于芦苇湖中。因“芦”与“胪”，“雁”与“宴”同音，故象征着“传胪赐宴”。传胪也成为了古代知识分子的统称，寓意金榜题名，前程似锦。

◉ The Ceremony for Announcing the Imperial Exam Result

The carving, *Ceremony for Announcing the Imperial Exam Result* is related to the imperial exam of *Keju* system in ancient China. The examinees who did the best in the imperial exam were divided into three *Jia*s in the Ming and Qing Dynasties. The first *Jia* composed of three examinees: the champion, who was called *Zhuangyuan*; the one in the second place, called *Bangyan*; and the one in the third place, called *Tanhua*. The second and the third *Jia*s both include several examinees respectively. Whoever got the first place in the second *Jia* was called *Chuanlu*.

When the day to announce the results of the imperial exam came, the emperor would announce the rankings of the examinees on the court by himself, and then the officials at the threshold of the court would pass on his announcements to the guards at the gate. The guards would shout together the names of those who were in *Jias* for six or seven times. This ceremony was called *chuanlu*, which means calling the roll of the top examinees, followed by a big feast given by the Emperor. The custom started from the Song Dynasty.

This carving shows three wild geese among the reeds. In Chinese, "wild geese" and "feast" are of the same pronunciation *yan*; "reed" has the same pronunciation with *lu*. So wild geese and reeds in the carving are to symbolize the ceremony for announcing the Imperial Exam result.

42. 鸡鸣戒旦 位置:钟楼牌坊下方东侧贴角花板

成语"鸡鸣戒旦"出自《诗经》。《诗经》是中国汉族文学史上最早的一部诗歌总集。

《诗经》收集了自西周初年至春秋时期这五百多年的三百零五篇诗歌,内容分风、雅、颂三部分。"风"是15个不同地区的地方民歌,共160首;"雅"主要是朝廷乐歌,分大雅和小雅,共105篇;"颂"主要是宗庙乐歌,共40首。"鸡鸣戒旦"出自《诗经·齐风·鸡鸣序》,即齐国的地方民歌:"鸡鸣,思贤妃也。哀公荒淫怠慢,故陈贤妃贞女夙夜警戒相成之道焉。""言古之贤妃御于君所,至于将旦之时,必告君曰:鸡既鸣矣,会朝之臣既已盈矣,欲令君早起而视朝也。"意为古代贤妃与君主同眠,到了鸡鸣之时,就提醒皇帝早起上朝处理国事。这是"鸡鸣戒旦"一词的典故。后人多用该词来形容晨起而作,因怕耽误正事,天没亮就起身;或形容过度紧张。《晋书·赵至传》载:"鸡鸣戒旦,则飘尔晨征。"何逊有诗云:"我为旬阳客,戒旦乃西游,君随春水驶,鸡鸣亦动舟。"

作品刻三只雄鸡于牡丹花圃,其中一只引颈鸣晨,神气活现。因"鸡"与"吉"谐音,而牡丹代表富贵,故该作品又名"吉祥富贵"。

◉ The Cockcrow at Daybreak

This carving is based on a story from the first Chinese collection of poems, *the Book of Songs*, which composes three parts: contemporary folksongs in fifteen different regions, songs in court and religious songs. This idiom is based on a folk song in the Qi State of the first part.

The idiom means to get up as soon as the cock crows, thus very early. It originated from a folk song about a virtuous imperial concubine who woke up her husband, the Emperor, as soon as the cock crowed so that he could get up early to handle the state affairs. The idiom was frequently used by people of later generations. A bibiology wrote, "The cock's call signifies the coming dawn, and an immediate morning journey is required." A poem read, "I am from Xunyang. I will travel west at cockcrow. You had better start your boat journey along the spring water."

In the picture three cocks of different poses are carved in the morning garden. One of them is crowing with its neck stretched.

43. 福寿图　位置:山门牌坊立柱和花板顶饰

"福寿图"表现了人们对幸福和长寿的祈盼。在雕刻、绘画、刺绣等艺术作品中,寓意幸福、长寿的意象除了寿桃、松鹤外,还有蝙蝠、寿字纹等。古代中国有"多子多福"之说,认为子孙多就意味着幸福。而葵花、石榴等植物多籽,与"多子"同音,故被认为象征多子多福。而蝙蝠因为与"遍福"谐音,寓意"遍地是福",也经常出现在艺术作品中。

作品上方刻寿字纹,下方刻着两朵葵花和一只立体蝙蝠,形成一幅精美的福寿图。其寓意为多子多福,福寿满堂。

◉ The Picture of Happiness and Longevity

This picture expresses the good wishes for happiness and longevity. Ancient Chinese thought peaches, pine trees, cranes and turtles were all symbols of blessing and longevity. The Chinese character *fu* means blessing and happiness. Sunflowers symbolize more children and blessings, for they have lots of seeds. In Chinese, "seed" and "children" are of the same pronunciation. Bat shares the same pronunciation with *bianfu* meaning full of blessings, so bat signifies "full of happiness and blessing."

With a three-dimentional bat surrounded by sunflowers, the carving implies the wish of more offsprings and more happiness.

44. 龙凤呈祥 位置:钟楼、鼓楼拱形门头上方

砖雕中拱形门头上的装饰是飞翔的龙凤和祥云。龙与凤是中国文化中最典型的祥瑞动物形象。《龙凤呈祥》象征吉祥、幸福,是古代建筑装饰中很常见的艺术形象。

周代《礼记·礼运》记载:“麟、凤、龟、龙谓之四灵。”“四灵”中除龟确实存在之外,麟、凤、龙都是由古代图腾演变而成的假想动物。龙在中国文化中是地位最高的一种神物,封建帝王常自称为“真龙天子”。凤被看作仁义道德和天下安宁的象征,是传说中“太阳之鸟”图腾的演变。中国古代哲学认为人类应顺应自然规律,重视天地相合、阴阳相合的观念。既然龙代表帝王,象征天和阳,就必须要有一种神物代表皇后,象征地和阴。于是凤就成了与龙相配的神鸟。“龙凤呈祥”迎合了中国文化的乾坤相合、阴阳相合的理念,表达了中华民族追求吉祥、幸福的美好愿望。

◉ Dragon and Phoenix Indicating Good Fortune

Dragon and phoenix indicates good fortune. Dragon and phoenix are both mythical animals symbolizing happiness in Chinese culture. They often show up together in architecture decorations.

The book *Rituals and Forms* in ancient Zhou Dynasty stated, “Unicorn, phoenix, turtle and dragon are four animals of intelligence.” Except the turtle, the other three are all mythical. Dragon was thought to be the symbol of Emperor, while phoenix stood for the queen. Therefore, we can often see dragon and phoenix together in the decorations of ancient buildings. This brisk carving indicates the traditional

Chinese idea that a match of a dragon and phoenix brings good luck, showing the desire for happiness and harmony.

鼓楼牌坊
Carvings on the Gate of Drum Tower

45. 寿比南山 位置:鼓楼牌坊上层花板(西)

“寿比南山”是中国文化中用来祝福老人长寿的词语。

“寿比南山”来自于一副对联:“福如东海长流水,寿比南山不老松”。南山即终南山,又名太乙山、中南山、周南山,简称南山,指的是秦岭中西起陕西眉县,东至西安蓝田县的一段山脉。

历史上,终南山与道教和修仙有着密切的关系。《小雅·天保》诗中就有用南山比喻长寿的诗句,如“南山之寿,不骞不崩”。传说我国历史上伟大的哲学家与思想家、道教创始人老子,在西游入秦时,曾在此讲授《道德经》。

此作品中刻着寿带(又作绶带)、鸟和松树,寓意为“寿比南山不老松”。

◉ **Live long like the Mountain of Zhongnan**

It is an idiom wishing one to live long as the Mountain of Zhongnan from a Chinese couplet, “May happiness flow in like water in the East Sea, and may one live long like a pine on the Zhongnan Mountain.”

In Chinese history, the Mountain of Zhongnan had close relationship with Taoism. There was a poem using the Zhongnan Mountain as a symbol of longevity. It was said that the founder of Taoism, Laozi, had once preached in the Zhongnan Mountain.

In the carving there are a ribbon, some birds and pines, representing the meaning longevity like the pines on Mountain of Zhongnan.

46. 凤凰戏牡丹 位置:鼓楼牌坊上层花板(东)

作品中刻着两只凤凰围绕着两朵盛开的牡丹。

凤凰是中国古代传说中的瑞鸟,擅长歌舞,是百鸟之王。凤凰被看作仁义道德和天下安宁的象征。古人邢丙疏曰:“孔子有圣德,故比孔子与凤。”《山海经·南次三经》云:“丹穴之山有鸟焉。其状如鸡,五彩而文,名曰凤凰。首文曰德,翼文曰义,背文曰礼,腹文曰信。是鸟也,饮食自然,自歌自舞,见则天下安宁。”凤凰是凤鸟的总称,其雄性为凤,雌性为凰。因此,旧时男子求婚被称作“凤求凰”,而凤凰双飞有夫妇恩爱之意。

牡丹是中国国花,有花王之称,其高贵美丽的姿容、雍容华贵的气质深得中国人的喜爱。凤凰和牡丹结合在一起象征荣华富贵、幸福美满,寓意繁荣昌盛、吉祥如意。

◉ **Phoenix and Peony**

A pair of phoenixes and two blooming peonies are carved in the picture.

The phoenix is said to be the mythical bird of auspiciousness and the queen of all birds. It is good at singing and dancing. It often represents benevolence and virtue in Chinese culture. Xing Binshu (an ancient Chinese writer) said, “Confucius can be compared with phoenix because of his virtue.” That is to say, phoenix indicated virtue and peace in ancient China. Male phoenix is called *feng* and female *huang* in Chinese. In ancient times, marriage proposal was also referred to as “*feng* proposes to *huang*” in literature. So the image of a pair of phoenixes also represents the love between a couple.

Peony is the Chinese national flower, which is loved and praised all over the nation for its nobility and glamour. Therefore, people put the flower together with phoenix to symbolize happiness, wealth and auspiciousness.

47. 三顾茅庐　位置:鼓楼牌坊大额枋

“三顾茅庐”是指东汉末年刘备三次到诸葛亮的住处请他出山辅佐自己的事件。该故事最早见于诸葛亮的《出师表》:“臣本布衣,躬耕于南阳”,“先帝不以臣卑鄙,猥自枉屈,三顾臣于草庐之中”,“咨臣以当世之事,由是感激”。《三国志》中对此仅有“凡三往,乃见”的简略记述。后来,《三国演义》对这个故事进行了深入的描写,使得“三顾茅庐”成为了形容“求贤若渴”的成语。

刘备(161 ~223 年),字玄德,涿郡涿县(今河北省)人。徐庶和司马徽向刘备推荐诸葛亮之后,刘备同关羽、张飞一起前往南阳卧龙岗(今河南省境内)求贤,未遇而归。数日后,三人又一次冒着风雪来到卧龙岗,诸葛亮仍没有回来。刘备就写了一封书信,让诸葛亮的弟弟诸葛均转交给他。不久后,刘、关、张三人第三次前往隆中,等候良久,终于见到诸葛亮一面。“大梦谁先觉,平生我自知,草堂春睡足,窗外日迟迟。”这首诗就是孔明当时吟诵的。刘皇叔三顾茅庐让他十分感动,终于答应出山,并为蜀国立下了汗马功劳。

砖雕中,右上方茅庐内的榻上睡着一人即诸葛亮,他的一双鞋子摆在床榻前。有一茶童小心翼翼地在煮茶,旁边有一个水缸、两只水桶。煮茶的茶壶、书斋中的桌椅和笔墨书籍清晰可见。榻的右前方有透刻的窗棱和雨搭,细如火柴的竹竿支撑着雨搭。左边三人便是刘备、关羽、张飞,中间背着手的人是孔明之弟诸葛均。

◉ Three Visits to the Thatched Cottage

This carving, *Three Visits to the Thatched Cottage*, tells a story of Liu Bei, the Emperor of Shu in the Three Kingdoms period in China. Liu Bei was said to be a ruler who valued and respected wise people, so he paid three visits to a thatched house where lived Zhuge Liang, a rare talent, to invite Zhuge to work for him. The anecdote was initially recorded in Zhuge Liang's *Memorial on Sending the Troops*, in which Zhuge expressed his gratitude for Liu's appreciation and invitation. And now, the idiom means “yearn for the wise and able men”.

Liu Bei (161 ~223), his another name being Xuan De, is from Zhuo county (in nowaday Hebei Province). Because of Xu Shu and Sima Hui's recommendation, Liu went to Wolonggang in Nanyang (in nowaday Henan Province) with Guan Yu and Zhang Fei to visit Zhuge Liang. They failed to see Zhuge at the first two visits. Some days later, they made the third visit on a snowy day. At last they saw Zhuge, who felt indebted to Liu's three visits and decided to work for Liu. In *Memorial on Sending the Troops*, he wrote, "I was originally a farmer in Nanyang, only concerned with my own safety, without any aspiration of being well-known or rich. His Late Majesty, overlooking thc commonness of my origin, condescended to visit me three times to my humble cot and consulted me on the latest events. His magnanimity moved me deeply, and I consented to do my utmost for him."

In the carving, one can see a tea boy making tea at the bedside. A teapot, a table, a chair, a writing brush, an ink stone, a water jar and two buckets can also be clearly seen. In addition, two thin trestles obliquely supporting the window shelter make the scene alive. A pair of Zhuge Liang's shoes is placed at the bedside, as if they were real.

48. 李娘娘住寒窑 位置:鼓楼牌坊小额枋

"李娘娘住寒窑"是一则民间传说故事,源于宋代宫廷轶事。它是由真实的历史人物、事件原型改编而成的。李娘娘是宋代皇帝宋仁宗的亲生母亲李氏。李氏在皇宫中地位不高,仁宗虽然是她亲生的,却被送给皇后刘氏抚养。直到李氏去世,仁宗皇帝都不知道自己的生母其实是她。刘后去世后,仁宗方知实情。当得知亲生母亲李宸妃是被刘后虐待而死的,他恼怒万分。

在民间传说中,宋仁宗归于刘太后名下之事,成为"狸猫换太子"的原型。"狸猫换太子"是清代石玉昆所著古典名著《三侠五义》第一回的著名段落,它是这样一段虚构的情节:宋真宗的嫔妃刘氏、李氏同时怀孕。为了争当正宫娘娘,工于心计的刘妃将李氏所生之子换成了一只剥了皮的狸猫，并污蔑李妃生下了妖孽。真宗怪罪下来，李妃逃往民间，刘妃遂被立为皇后。宋真宗死后，李妃所生男婴，即后来的宋仁宗即位。李娘娘来到民间后，太监陈琳托秦凤将她送往陈州的家中。秦凤死后，娘娘便住进了寒窑。当时范宗华遵从父亲范胜的遗嘱，与李娘娘以母子相称，并搭棚住在娘娘的窑边伺候。娘娘因思念储君（皇太子）终日哭泣而双目失明。一日，包公陈州放粮回京至天齐庙，宗华便禀告其母。娘娘闻之，即去告状。包公问："你有何冤枉？"娘娘从内衣里掏出一个小布包，包公打开一看，布包内是一粒金丸，上刻"玉宸宫"三字和娘娘的名字。在包拯的帮助下，仁宗得知真相，与双目失明的生母李妃相见，而身为皇太后的刘氏则畏罪自杀。

本作品上刻的场景正是李娘娘拜见包公告状的一幕。图左寒窑前站立者是李娘娘，中间是包公，其他人分别是王朝、马汉等。娘娘面前下跪者是范宗华。

◉ **Imperial Concubine Li in a Humble Dwelling**

The carving tells about a folklore based on a real story in the Song Dynasty. Imperial Concubine Li was the mother of Emperor Ren of Song. But because of her humble status in the court, her son was taken away from her and was brought up by the Queen. Then Li was driven out of the court, though survived from a murder. However the son didn't know who his real mother was until the death of his step mother, the queen.

However, in the folklore told by a novel named *Three Swordsmen and Five Benevolences*, the plot turned out to have a happy ending: Liu the Queen framed Li by accusing her of giving birth to a palm civet, and even wanted to murder her. Then Li was saved by Chen Lin who entrusted Qin Feng to send her to live with a family in Chenzhou. After Qin's death, she lived in a humble cave-dwelling, with Fan Zonghua protecting her. Imperial Concubine Li and Fan addressed each other mother and son according to the last will of Fan's father. She went blind because she cried all day missing her son, the Emperor at that time. One day, Bao Gong (an honest and upright official in the Song Dynasty) returned to the capital, and arrived at Tianqi Temple. Fan told her the news; she went to Bao Gong. Bao asked: "What complaints do you have?" She took out a golden ball, with the name of herself and the palace she lived in on it. Bao Gong then reported her issue to the Empire, and finally she got together with her son, while the former Queen who framed Li suicided for fear of the punishment.

On the left of the picture is Imperial Concubine Li standing before her humble cave-dwelling. In the middle is Bao Gong the official. The person who knelt before her is Fan Zonghua. The carving implies that your kindness will pay you back one day.

49. 燕山教子 位置:匾额《鼓楼》西侧兜肚

燕山,姓窦,名禹钧,北京幽洲人。因居住在燕山(今北京西南),故称窦燕山。

传说,窦燕山少时行恶,30 岁了还没有子嗣。其父托梦给他让他“速悔过迁善”,“改过呈祥”。他将梦中父亲之言谨记在心,一改从前的恶行,开始周济穷人。他后来官至太常少卿,右谏议大夫。窦家高义笃行,家法严格,家庭和睦,是当时人们的表率。他的五个儿子都考上了甲科。《宋史·窦仪传》记载:“宋代窦禹钧五子相继及第,长子曰仪,为礼部尚书;次子曰俨,为礼部侍郎;三曰侃,为补阙;四子曰称,为谏议大夫;五子曰僖,为起居郎。”故称“五子登科”。后有侍郎冯道赠诗云:“燕山窦十郎,教子以义方。灵椿一株老,丹桂五枝芳。”《三字经》中以“窦燕山,有义方,教五子,名俱扬”的诗句歌颂此事。从此便有了“五子登科”的成语,常用作祝福语,寄托了人们对后代的期望。

这幅砖雕中,书案后端坐着的一位老者即窦燕山。作品的寓意是,一个人要多行善事,就会有善报。

◉ **Yanshan Educating His Sons**

Yanshan's original name was Dou Yujun. He was born in Youzhou (in nowaday Beijing), near a mountain named Yanshan, hence he was also called Dou Yanshan.

Dou Yanshan was said to be a villain when young. At age 30, he was still childless. In his dream, he was warned to stop idling and evil doings by his father, whose words never left him since then. So he began to help the poor, hoping his offsprings would be benefited from his benevolent doings. Later he had five sons and all of them passed the imperial exams one after another. His oldest son Douyi became an official in charge of court etiquette and imperial exams; his second son, Douyan, vice minister of education department; the third son, also named Douyi, a counselor; the fourth son, Doucheng, also a counselor; the fifth son, Douxi, an official in charge of recording the emperor's actions and words. Later the Dou family was often praised by the later generations. They became well-known throughout the history. They set an example for family education in ancient China.

In this picture, the old man sitting behind the table was Dou Yanshan. It suggests that people who devote themselves to charity will be rewarded for that.

50. 王质烂柯　位置:匾额《鼓楼》东侧兜肚

"王质烂柯"的典故出自中国古代神话故事。王质,樵夫名。柯,斧柄。烂柯,斧柄烂尽。

据南朝梁任昉《述异记》上卷记载,信安郡石室山,昔时有一樵夫名王质,衢州人。一日上山伐木,至,见二老者围棋,质因视之。老者将一物似枣核赐予质,让质含之,不觉饥,一棋下完,老者谓质曰:"何不去?"质起视,斧柯尽烂。至家众云:已数百年,亲人无复存者。《述异记》后,遂以"烂柯"作为围棋的别称。一说王质复如山得道,人往往见之,因名其山曰"烂柯山"。故烂柯山被传是中国围棋之根,围棋界亦用"烂柯"来象征醉心棋艺。烂柯山在春秋时期被称为石室山,为姑篾国一大胜地。晋朝中期,王质的故事广为流传。北魏时期称该山为悬室坂,唐初称之为石桥山。元和初(约806年),此山开始被称作烂柯山。

作品中雕刻着两位对弈的老者。他们前方站着一个扛着大斧的人就是王质。"烂柯"比喻世事的沧桑巨变,正所谓"阅世深疑已烂柯"。

◉ **Wang Zhi's Rotten Ax Handle**

It is an ancient myth said to have happened in Shishi Mountain in Xinyang.

According to the book *Record of Strange Things* in the Nan Dynasty, once there was a woodcutter named Wang Zhi from Quzhou. One day, he climbed the mountain to get some wood. When he got there, he found two old men playing the game of go. He stopped to watch it. One of the old men gave him what looked like a date pit as food. He ate it and didn't feel hungry any more. When the game was over, the old men asked him, "Why don't you go home?" When turning around, he found his ax handle rotten. When he hurried back home, he found himself in a completely strange place. The villagers told him that several hundred years had passed since he left home. Rotten ax handle became another name of the game of go in China from then on. Another saying is that Wang Zhi went back to the mountain and became an immortal, so the mountain was called the Rotten Handle Mountain.

The carving shows two old men playing the go game, and Wang Zhi standing beside them with a big ax on his shoulder. It means that the world is changing rapidly ahead of one's awareness. And it implies that "Time and tide wait for nobody".

51. 喜鹊登梅 位置:鼓楼拱门上方西侧贴角花板

该典故的产生和中国人对梅花的喜爱有关。梅花开在深冬和初春,是一年里百花之中开得最早的花,又称"报春花"。故有中国对联"春为一岁首,梅占百花魁"。此联是春节时常见的咏梅时令联,联首两字是春梅。在中国文人笔下,梅、兰、竹、菊因其高雅的品格被称为花中"四君子",寓意着人的品格高尚。

喜鹊被认为是吉祥鸟,有"报喜鸟"的别称。自古就有"鹊鸣兆喜"的说法。古时候,人们如果听到喜鹊飞到家门口的树上欢快地鸣叫,就认为这预示着家中喜事将至。"喜鹊登梅幸福来",喜鹊登上梅梢与"喜上眉梢"谐音。两只喜鹊中间加一枚古钱则寓意着"喜在眼前"。《开元天宝遗事》记载:"时人之家,闻鹊声皆以为喜兆,故谓喜鹊报喜。"《禽经》也记载了"灵鹊兆喜"之说。

此幅砖雕作品刻一对喜鹊落在一株怒放的梅花枝头。梅花仿佛暗香浮动,两只鸟儿一鸣一啄,十分有趣。"喜鹊登梅"又称"梅占花魁"、"喜鹊闹梅",极具喜庆之意。

◉ Plum Blossoms and Magpies

Plum blossoms bloom the earliest among all flowers in a year, telling the coming of spring, thus is called "Chinese primrose". There is a Chinese Spring Festival couplet saying, "Spring is the first season of the year, while plum blossom is the first one to bloom in a year." In Chinese culture, plum blossoms, orchid, bamboo and chrysanthemum are regarded as elegant flowers, signifying noble people.

Magpie is thought to be a bird of luck. Its name means a bird telling good news. Back in old times, if someone heard a magpie chirping joyously on the tree in front of his house, he would think that fortune would come to his family. A magpie in Chinese is pronounced as *xique*, meaning the bird of happiness, and plum blossoms *mei*, meaning one's brows. So together they mean happiness is on one's brow.

In this carving, on the plum tree with flowers blooming, there is a pair of magpies. One of them is chirping, and the other is pecking. The plum blossoms look as if they were real. The carving indicates happiness and joy coming with the spring.

52. 鸳鸯戏莲 位置:鼓楼拱门上方东侧贴角花板

鸳鸯是一种雄雌偶居不离、生死相依的鸟。崔豹的《古今注》中说:"鸳鸯、水鸟、凫类,雌雄未尝相离,人得其一,则一者相思死,故谓之匹鸟。"鸳鸯常用来比喻忠贞的爱情和成双成对的事物。莲的别名是"水芙蓉",比喻纯洁美丽的女子。莲的果实是莲子,寓意着"连生贵子"。

在中国古代,最早是把兄弟比作鸳鸯的。《文选》中有"昔为鸳和鸯,今为参与商","骨肉缘枝叶"等诗句,这是一首兄弟之间赠别的诗。晋人郑丰的《鸳鸯》在序文中说:"鸳鸯,美贤也,有贤者二人,双飞东岳。"比喻陆机、陆远两兄弟。三国时期曹植的《释思赋》:"况同生之义绝,重背亲而为疏。乐鸳鸯之同池,羡比翼之共林。"最早将鸳鸯比作夫妻的是唐代诗人卢照邻的《长安古意》中"得成比目何辞死,愿做鸳鸯不羡仙"的诗句。李时珍的《本草纲目》中形容鸳鸯:"终日并游,有宛在水中央之意也。或曰:雄鸣曰鸳,雌鸣曰鸯。"

作品中,一对鸳鸯俯瞰着两朵硕大美丽的水芙蓉,寓意着夫妻恩爱,白头偕老。

◉ Mandarin Duck and Lotus Flower

Mandarin ducks always appear in pairs. In *Selections of both Past and Present* by Cui Bao, he said, "Mandarin ducks live in the streams of mountain areas. They always show up in pairs, thus they were also called *Piniao* (paired birds)." So in Chinese culture, mandarin ducks often refer to faithful lovers. Lotus flowers often represent a virgin and pretty female.

In ancient China, people originally employ the image of mandarin ducks to refer to brothers. Later they became a symbol of lovers. Li Shizhen, an expert of traditional Chinese medicine in the Ming Dynasty, said in *Compendium of Materia Medica*, "A pair of mandarin ducks swam intimately in the water. One is male, and the other is female."

In the carving, a pair of mandarin ducks and beautiful lotus in the rippling water indicate that a devoted couple will live in harmony and grow old together.

注:本书介绍的52幅砖雕作品均选自亳州花戏楼山门牌坊,其中选自正门牌坊的有35幅,钟楼牌坊9幅,鼓楼牌坊8幅。本书中的砖雕图名称系参考魏彪所著《花戏楼的砖雕艺术》一书。

后　记

中华文化博大精深，砖雕艺术精美绝伦。本书从一个侧面对亳州花戏楼精美的砖雕所蕴含的精深文化进行解读，并尝试将其翻译成英文。由于作者的文化水平、翻译能力及时间有限，可能在对花戏楼砖雕文化的理解和翻译的精确度方面还存在不足，在此恳请专家和读者批评指正。

本书是作者主持的安徽省高校人文社会科学课题“亳州花戏楼砖雕艺术文化解读及英语译介研究”的主要成果。砖雕艺术文化的研究和汉译英翻译的实践，同时也提高了本人的学术研究能力和文化审美自觉性。本书的出版离不开各方的支持。首先，作者要感谢安徽师范大学的孙胜忠教授、阜阳师范学院的梁亚平教授在课题立项中的建议和指导；感谢张辉等课题组主要成员的参与和帮助；感谢上海理工大学的贾从永先生、杭州国际语言学校的张倩女士对英文译文的校对和勘误；感谢亳州花戏楼文物管理处和亳州文联提供的支持。在此，作者也衷心地向所有在本书编写和翻译过程中提供帮助的人一并致谢。最后，感谢学校和出版社的支持。正是有了上述单位的帮助和个人的努力，本书才得以问世，让花戏楼的砖雕艺术和文化意蕴以汉英双语的形式得以展现。

希望本书能带给读者一丝新意和愉悦。

唐利平

于亳州水木清华园

2014 年 2 月 24 日